ON THE
Dual Reality
OF EXISTENCE

BERT WARREN, PH.D

On the Dual Reality of Existence
Text Copyright © 2013 by Bert Warren
Santa Fe, New Mexico, U.S.A.

Published by Decision Point Publications at
Smashwords
Santa Fe, New Mexico 87508

All rights reserved. No part of this book or its website may be reproduced, stored, or transmitted in any form or by any method, electronic or mechanical, including photocopy, scanning, recording or any information retrieval system without permission in written form from the author.

Scripture quotations are from The Revised Standard Version of the Bible copyright © 1946, 1952 and 1971 by the Division of Christian Education of the National Council of Churches in the USA. Used by permission. All Rights Reserved.

Permission granted by representative of spiritdaily.com to use factual material from www.spiritdaily.net/clark.htm relating to Maria's OBE and Kimberly Sharp Clark.

ISBN: 978-1482370904

Book design by Maureen Cutajar
www.gopublished.com

Acknowledgements

This text is dedicated to those imaginative and creative individuals who differentiated us from other vertebrates by modifying and accelerating our ascent from natural evolution to technological evolution. From learning to use fire to space travel and in all probability to an incredible future (if we do not destroy ourselves first). I therefore gratefully acknowledge the labor and imagination of all of those creative individuals without whom our civilization and this work would not be possible.

Contents

Introduction 1
Chapter One: What Is The Truth 5
Chapter Two: Understanding 11
Chapter Three: The Gift of Life 17
Chapter Four: The True Nature of Death 21
Chapter Five: Existing Evidence 25
Chapter Six: Personal Experiences 27
Chapter Seven: Crop Circles 35
Chapter Eight: Animal Mutilations 45
Chapter Nine: Unidentified Flying Objects 49
Chapter Ten: Remote Viewing 57
Chapter Eleven: Out of Body Experiences 61
Chapter Twelve: Reincarnation 69
Chapter Thirteen: Near Death Experiences 77
Chapter Fourteen: Possibilities 85
Chapter Fifteen: The Holographic Universe 89
Chapter Sixteen: The Simulation Hypothesis 99
Chapter Seventeen: Dual Reality 103
Chapter Eighteen: Other Possibilities 123
Chapter Nineteen: Remaining Questions 127
Chapter Twenty: Our Fate 133
Addendum 135
Summary 141
Final Conclusion 143

Introduction

This IS NOT a theology text. The existence or nonexistence of a supreme being will be addressed, but only briefly. However, other basic spiritual questions are asked and answers provided; questions such as do souls exist? And here is the answer to that question: YES, an emphatic YES based on facts that will be provided. Other metaphysical questions are examined that you may have never even considered, such as how and why souls came into existence, why reincarnation exists and why life is an adventure. Open your mind and enjoy the trip.

Is death final? Does it end our existence or do we continue on to another level of reality? These are the ultimate questions and in an attempt to answer them we must delve into the nature of both life and death. We must also understand what is currently believed to be the nature of our universe. These determinations are absolutely essential for us to attempt to answer the above questions. Each of these topics is complex, consisting of a fabric of data and surmise that deserve a library of texts. Clearly therefore an in depth consideration of each of these topics will not be covered here nor is that necessary. Only key reported events from each paranormal topic and their relevance to the existence of the dual reality we live in will be presented. And as you will note, all conclusions derived from these events will be based on factual and confirmed information. *(If desired, each of these phenomena can be studied in greater detail in the many excellent books and articles on these topics).*

And as you will find, we shall use this key extracted information to go several significant steps further into the question of spiritual existence than is provided by texts specializing in one specific area of investigation.

Our findings will be used to establish a new and uniquely unified hypothesis defining the nature of life, death and the dual reality of existence. We shall further propose a possible answer as to <u>how</u> this came to be.

We shall also provide a possible answer as to why this dual reality of physical life, death and immortal spiritual existence is present. You will find the proposed answer to be startling and totally different from any you have heard before. So then, open your mind and let's get started building evidence.

The initial section of this text summarizes our understanding of the universe, the miracle of life and my own personal and essentially unexplainable experiences. Then unresolved phenomena are explored such as crop circles, animal mutilations, UFOs and remote viewing.

The purpose of these reviews (this walk) is to bring clearly into light how much we know and how much more we still have to understand.

Hopefully, this process will provide the openness required as we then explore the nature of life, death, consciousness, memory, reality and spirituality (here we jog).

Our next focus will introduce projected technological advances that will lead to immortality (now we run).

Finally (we fly), a new concept will be proposed advancing the proposition of a dual reality that defines the how and why of the nature of our existence and immortality.

On the Dual Reality of Existence

Nothing in this text is intended to denigrate anyone else's paranormal experiences. The difference here is that the cases cited have been confirmed by undeniable evidence. As a scientist I personally feel compelled to define truth to what is observable and repeatable fact; and even that is in question at times. Nonetheless this is what I will base my premise on as to what truth is.

Chapter One
_____

What Is The Truth

Please bear with me as I define what to me and probably to you as well is simplistic and obvious, but to define truth, its premise must be spelled out.

We live in a universe that for all extents and purposes is infinite. Possibly there was a beginning for the universe as we know it but undoubtedly infinite space and energy always has and always will exist. So, is our universe finite; most probably not, as stunning as that conclusion may be. We live in a universe that has always existed and always will exist in one form or another.

In this universe there is an incredible variety of matter and energy. We are most familiar with matter. To direct our attention to us we will leave the exquisitely proposed important details of how the universe, matter and energy came to be and attempt to directly focus on the creation of humanity.

We live in a universe 14.6 billion years old, consisting of over 100 billion galaxies with each containing over 100 billion stars. One of those stars we call "The Sun." We believe that we know how stars and their satellites were formed. We believe we know not only how stars are born, but also how they die. We know as a fact that we live on the 3rd satellite from the Sun that we call Earth. We know that life was created on Earth and as far as we know human life had eventually evolved from over the 4.54 billion years of Earth's existence. At this point, one may well ask if this has been a result of evolution or divine intervention.

Science has clearly demonstrated the step-by-step process of evolution from microorganism to man. Just the commonality of DNA alone is undeniable, not to mention the basic metabolic cycles and stepwise organ and functional similarities as we move up the animal chain. Were these changes random or intelligently designed? The truth is that there is no absolute evidence ruling out either direction. Did critical changes in DNA occur resulting from random hits of radiation, fortunate metabolic "errors," alien intervention, or the hand of God?

On the Dual Reality of Existence

We cannot be certain, but as we cannot prove alien intervention or the hand of God we are left with the most feasible explanation; random change, evolution, survival of the fittest. Absolute truth, no, but the most likely of the possibilities.

It is clear that humans are animals, mammals to be more or less specific. I am certain that I do not have to go into blood and guts here to prove that; it's obvious. But humans are unique. Bit by bit by using our developing superior brain, unique voice box, and the gift of upright posture freeing our hands we have evolved and continue to evolve into becoming a superior life form.

Slowly at first, instinct provided group living, hunting and defense. Cleverly we designed tools to cut meat and pierce prey. Progress continued; animal skins for warmth, primitive language, clay vessels to hold water, fire for warmth and then to cook food, also the domestication of animals for a reliable food source, hunting and transportation. This followed by the discovery that desirable vegetation could be cultivated.

All this gave our kind tremendous survivability advantages and our population grew and spread across the globe. It then became apparent that different regions produced or had available materials that were desirable and could be traded.

Trade required accounting leading to the development of numbers and writing. Seasonality became more important for planting and the understanding of seasons provided foresight as to weather expectations. Monitoring of seasons was learned by tracking the sun and stars across the heavens; the significance of time became apparent.

So, step-by-step, across the millennia we separated ourselves from other creatures and evolved to form complex civilizations. From logs to canoes we navigated rivers; from logs to wheels we greatly enhanced transportation. We clearly became superior beings with our large organized civilizations, ships and caravans for trade, parchment for writing and societies with civil command and control.

Still there was a long way to go to arrive at where we are now. Clearly we are currently advancing asymptotically; electricity, energy sources, mass production, automobiles, aircraft, refrigeration, radio, television, computers, cell phones, medicine, space travel, to name a few advances. We are what we are as a result of technology. Just take out one item and see what life would be like. Can you envision a society without clothing? A string quartet on stage performing nude? Your boss without clothes? And that is just one "unessential" change.

On the Dual Reality of Existence

Yes, we are what we are as a function of technology. We did this ourselves and will continue to modify our nature as a result of our efforts.

But to be exact, who are we? Many, many humans exist and have existed, but how many contributed to how we live now and how many will contribute to how we will live in the future? The majority of us are just "along for the ride" enjoying the fruits art, science and technology have and will provide. However although only some of us push back at the frontiers of knowledge, all of us enjoy its results. How far will we get? Space travel to new civilizations? Immortality?

Frankly, at this point, anything of practical significance that is feasible can and probably will happen. Immortality is not impossible and it is quite probable that technology will get us there.

Chapter Two

Understanding

If we were dealing with absolute knowledge and complete solutions, answers to all our questions would be certain. If, on the other hand, we are dealing with concepts that are continuously being refined and challenged we do not have certainty and many questions will remain to be answered. As the latter is indeed the case, new hypotheses and determinations defining the nature of matter and energy, materialism and spirituality, and life and death are to be expected, have occurred and will continue to occur.

Once, we were the center of the universe and the sun rotated around the earth. Once, witches brought on plagues and bad air (malaria) and gods determined the fate of battles, all peoples and all creatures in the world.

However what we believe now, is very different. So then, what do we favor currently? There are now two basic approaches to what is considered to be reality, faith based and scientifically based.

The faith-based belief is in a supreme and ever-existing spiritual being controlling everything. Unfortunately this approach is not based on firm, repeatable evidence. (For this reason a discussion of religious beliefs throughout the ages will not be presented here *(for this, see Death and After by Annie Besant Theosophical Publishing Society 1906 and many other more recent texts)*.

Nonetheless, a religious approach to reality still exists in the minds of many or most of us. Science, on the other hand, prefers to define reality based upon material, repeatable and ideally mathematically based fact. Unproven data are only accepted for consideration as hypotheses or theories.

Of all the prevalent scientifically based theories, the one most exciting to me is the attempt to find a unified theory for everything. Currently the favored scientific constructs for the nature of energy, matter and the universe are based on superstring theory, ten or more dimensions and the existence of multiverses.

On the Dual Reality of Existence

Let us briefly examine them.

Once the basic unit of matter was the atom. The number of neutrons and protons in its nucleus and the number of electrons spinning around it defined the specific nature of each atom. Then newer smaller particles were discovered, quarks, bosons, leptons, hadrons and so on. This concept of sub-atomic particles has not been abandoned but is being supplemented by units of matter so small that the boundary between matter and energy is being blurred (or is it being better understood)? Thus we now have string theory.

In string theory the basic "particle" is an incredibly small string, the oscillations and entanglements of which determine the exact nature of each specific string. Actually, however, strings are not believed to be particles, but rather they are conceived of as being packets of vibrating energy. These incredibly minute packets, as integrated groups, along with bosons, are defined as the basic units of what gives matter form and provides for energy fields. These fields in turn then define everything in the universe. Quite a seemingly wild theory, but thus far it holds up when challenged.

So there you have it, space-time, suns, planets, people and bugs all consist of vibrating strings. Difficult to believe, but again, it seems to hold up when challenged – thus far! As for multiverses and ten or more dimensions, what can we know about dimensions we cannot sense or universes we cannot comprehend?

So what are we, superstrings, and how does this affect our perception of reality? Here is a construct I conceived. Let's start with nothing – no matter, no energy, no strings, no space ... just nothing. Think about it, because when you do you will realize that absolute nothingness is utterly impossible! Why? Because this theoretical nothingness must be surrounded by space and space is not nothing, it is something.

How much space? An infinite amount! At this point one may well ask what is keeping space from collapsing – what is, in a sense, holding it up? This requires another key ingredient: energy. By necessity therefore energy also always has existed. Here we are clearly defining total and eternal infinity where space and energy can, does and must exist.

Is this energy the basis for the formation of strings and are strings the basis of matter? Do strings attract each other to form larger aggregates of matter? Do larger aggregates of matter coalesce further to produce very large masses of matter? Do these large masses attract each other to form an extremely small but incredibly dense single mass? Does this mass reach critical levels and then explode outwards in a big bang? Are strings and superstrings the basis for the formation of everything?

According to currently accepted theory they are. If we consist of superstrings by what miracle is their nature harnessed to form living reproducible creatures? Will this theory hold up? I do not know. But what is very clear is that the basic nature of our existence, our lives, our surroundings and our understanding of the universe still leaves much to be desired.

Therefore we must remain open to considering new possibilities, and as best we can determine, such possibilities should be based on reproducible evidence. However, where there is no evidence, hypotheses should be considered and such concepts should not be ruled in or ruled out until confirming evidence is accrued one way or the other.

So there we have it, a materialistic theory of everything. But what about spirituality? The obvious difficulty with spirituality is that it is not material. Material items can be seen, they can be touched, they can be smelled, they can be manipulated and they respond to consistent actions in a consistent manner. Spiritual "items," as far as we know, can do none of these things in a repeatable consistent manner. Thus the question, are spiritual experiences real and if so how can they be proven?

As an article of faith, this is simple. "I believe" covers it all – but belief is not science, it is faith and faith is yet to be factually and repeatably proven. In our current theoretically based belief systems of quantum mechanics, string theory, ten dimensions and multiverses, just about anything is possible, even spirituality. We, however, do not live in ten dimensions; we live in four dimensions consisting of length, width, height and time. Other dimensions and universes are not available to our senses. Is it then possible that spiritual worlds exist that we cannot sense? So then, are spiritual (gestalt) entities and existences possible? Well, apparently they cannot be ruled out and they cannot be completely proven as, <u>presumably</u>, there is no totally indisputable evidence for coming to either conclusion.

As best we can, however, we will explore reported unusual and spiritual happenings and attempt to sort out what seems to be possible, what seems to be probable and what is certain.

Chapter Three

The Gift of Life

Without question life is a gift. Without question it is a miracle. A miracle based on so many factors being exactly suitable for our existence, that one might consider the presence of a supreme life force as being an absolute requisite.

Here, however, we have the problem of Atlas holding up the earth. Who then is holding up Atlas? A woman who believed that Atlas indeed held up the earth was asked: who is holding up the earth? She replied, "Atlas". Then she was asked who is holding up Atlas? And she replied, "It's Atlas all the way down!"

In Exodus 3:1 to 3:15RSV when Moses came to the "Mountain of God" in Midian and while tending a flock, "an angel of God" appeared to him in a flame of fire out of the middle of a bush: and when he looked at it, he saw that the bush burned with fire, but the bush was not consumed. And when God saw (the bible does not state earlier that God's presence was in the bush) that Moses had noticed this, God called to him from out of the middle of the bush, and said, "MOSES, MOSES I AM HERE".... And Moses said to God, "behold, when I tell the people of Israel about this, and say that the God of your fathers has sent me to you they will ask, what is his name? Then, what shall I say to them?" And God said to Moses, "I AM WHO I AM": and He said, "THUS SHALL YOU SAY TO THE CHILDREN OF ISRAEL, I AM HAS SENT ME TO YOU."

This is a perfect answer for a God who has always existed. But a burning bush unconsumed, an angel of God within the bush, God speaking from within a burning bush? Absolutely ideal for ancient alien enthusiasts! Nonetheless the answer given by the entity, whether God, alien visitor, time traveler, or myth "I am who I am" perfectly describes the existence of a being without beginning or end. And even if you question and exclude the existence of such a being, the completely staggering fact of the always-existent reality in which we live is undeniably true.

So here we have it, infinite space, infinite time,

On the Dual Reality of Existence

and ever-existing energy, as strings, in some way creates the universe with its myriad galaxies, stars, planets, moons, asteroids, life, everything, all based on scientific theory.

And this only the beginning; black holes, quasars, dark matter, Higgs bosons, supernovas, on and on. Supernovas as we do know are the reason for the formation of the heavier elements essential for life. Think of the miracles. Our planet in the goldilocks zone, not too hot or too cold. What if hydrogen did not combine with oxygen to make water? What if carbon complexing with other molecules to form long and varied chains did not occur? What if the earth did not rock back and forth? What if there was no moon?

Well, fortunately for us all these phenomena do occur and did occur in such a manner, and some in increasing complexity, so as to allow the formation of living creatures and eventually to us. Life consisting of molecules constantly modifying and forming cells, each formed cell a complex of protoplasm, mitochondria, RNA, DNA, membranes and cycling metabolism. Each molecule with a specific function, each organizing into cells, organs, muscles bones and blood. The blood produced pumping through arteries and veins, oxygen and nutrition carried throughout an integrated whole to produce functioning entities each defined by their individual genetic codes.

Thus we have entire organisms with the ability to replicate, evolving into animals with the ability to think and culminating with humans having the ability to create. All dwelling in a planet perfectly situated with water and oxygen and a moderate climate ideal for life. And as you know, this is an incredibly oversimplified exposition of how life came to be and what life is.

(So we take this unbelievable gift and not only destroy life to live but destroy life for beliefs. Beliefs that cannot be proven, beliefs taken to be true as an article of faith, beliefs lacking the firm evidence required by science, beliefs that can and induce prejudice, hate and murder.)

Still, what I find to be most amazing is our belief that EVERYTHING consists of strings. So then, what is life? How can we define it? To me the answer is clear. Life is an absolutely incredible miracle. Think of all the steps required for you to have come into existence, from galactic to microscopic. It is a gift, a gift of chance, a gift of design or a gift from a supreme intelligence. As recipients of this gift we are granted an unfolding adventure from birth to death: love and loneliness, illness and health, peace and war, luxury and poverty. We all face these possibilities and more, we all participate in this adventure.

Chapter Four

The True Nature of Death

What is the true nature of our existence after mortal death? This question has never been answered to everyone's satisfaction and perhaps never will be answered. Maybe there is a heaven rewarding the good and a hell punishing the evil. Maybe there is nothingness in the sense of a total lack of conscious awareness. Maybe there is an existence of which we are totally unaware or maybe there is an existence from which we can only capture fleeting nebulous glimpses.

Do our religious leaders know and present us with the truth? Ask them they will tell you. But if so, which religious leaders; Catholic, Protestant, Jewish, Buddhist, Moslem, Hindu, and so on? The simple fact is that their defined "truths" differ between them. And this, unfortunately, clearly indicates that the absolute truth is not known by them.

As I am sure you are aware, there are other alternative possibilities in regard to the true nature of death aside from religious teachings; alternative teachings that are supported by some absolute and irrefutable facts.

Now what our eyes tell us and what living tells us is that we are born, we live and then we die. If we are cut we bleed. If we are hungry or thirsty we eat and drink. If we are shot or stabbed or seriously ill, we may die. Cold, wet, lonely and so on, we know what happens. These are clear facts, they are real, they are apparent. We share these truths with all higher animals and all life as we know it. And as we know, all living creatures die. What else do we share with other creatures? Paramecia, single cell animals, have directed motion, eat, eliminate waste, can have sexual reproduction (conjugate) and flee from threats. They have no brain and no nervous system but still they live, multiply and react; they are alive. So then, what is cognition? In "higher" animals, hearts pump blood through arteries and veins, each has blood cells, each livers, kidneys, intestines, brains and so on. We all consume food, we all breathe, have sex to procreate, eliminate waste and on and on. So why should we be different in terms of death? Read on!

So what is the truth about our fate? Which belief is correct? Do we simply disintegrate in both body and soul and spend the rest of eternity as insensate disorganized organic and inorganic molecules or atoms, or are we reduced to incredibly small quantum bits? At one with the universe? Maybe ultimately as infinitesimal small randomly vibrating strings, or are we reborn, or bound for heaven or hell?

In terms of absolute certainty, we simply do not know. There is no available science to unravel this question and define an answer. There is however existing evidence from which one can extrapolate and speculate.

TRUE REALITY:
THE HUNDRED YEAR GAME OF LIFE

We are here, but what does that mean? We live, have experiences and eventually die. But is this really all there is? Is this true reality? Is this what it is all about, the bloodshed, the violence, the boredom, the glory, love, faith, death? Does a caring God destroy a life, a child, a people, a nation, a race, a species? Must we kill; to eat, for nourishment, for pleasure, for hate, for glory, for beliefs, for survival? Is this what life is all

On the Dual Reality of Existence

about? Existing? Perhaps it is about studying our nature or studying our surroundings, appreciating love and beauty or appreciating the creations of a supreme being.

On the other hand, is it just a sequence of random acts of pleasure and joy followed by horrific experiences each in an unpredictable sequence? Is it just an accident of physics, chemistry, birth and evolution; temporary life with permanent insensate death? I do not believe so. As you will see, there exists evidence supporting our continuing survival after the death of our bodies. What I believe probable is that living is about immortality, adventure, and strange as it might seem, (and as you will learn), escape from forever.

Chapter Five

Existing Evidence

Perhaps by studying existing evidence the underlying nature of our apparent short and unpredictable existence can be determined; this is what we will attempt to do.

To accomplish this end, however, we must first enter a world of contested concepts and challenged evidence, which encompass both material and spiritual (paranormal) phenomena. Material occurrences that are disputed include the cause and appearance of crop circles, the reality or non-reality of UFOs and the nature and cause of animal mutilations.

Contested paranormal experiences pertinent here are claims and discussions relating to the validity of remote viewing, reincarnation and out of body experiences. We must also examine contested phenomena closely related to the nature of death (near death experiences). These contested phenomena must be studied to see what they can teach us about the reality of spiritual existence. However, only essential facts will be presented to avoid the possibility of infringing on copyrighted material.

Chapter Six

Personal Experiences

Personal experiences are just that. They occur, are not scientifically based and may or may not be supported by witnesses. Significantly, they are absent of hard evidence, cannot be demonstrated and cannot be repeated. Some, however, undoubtedly have happened, thus demonstrating again that there are unresolved questions requiring answers.

Some paranormal events I have experienced are clearly inconsistent with reality; but be that as it may, the events I experienced and will describe have been witnessed and are true. They are presented because they are inconsistent with reality as we know it and as such, they open us to the consideration of new possibilities.

But believe them or not, they do not invalidate the confirmed facts that will be pointed out in this text.

Others, aside from myself have had unexpected and inexplicable experiences; there is no question but that I am not alone in having them. Talking with others usually draws forth experiences they have had that cannot be explained in terms of material reality. Neither they nor I claim frequent experiences of unexplainable events. However most can describe several such experiences. I have had four or five in a lifetime of over 82 years thus far. Most others who I have spoken to about this also have experienced several inexplicable occurrences. I assume that most all of us have had them as well. So what is going on? We do not know, but we can try to explain these events. Clearly they indicate that there is more to existence than meets the eye. (But we know that already).

What about you, have you had inexplicable visitations; seen UFO's, or experienced other phenomena and occurrences that defy accepted logic? Have you seen someone disappear with your own eyes or noticed subtle but real changes in your environment that cannot be explained? Have you had other things happen to you that no one can logically explain by our current concept of reality? In all probability you have.

By way of example, has a past companion or loved

On the Dual Reality of Existence

one recently deceased seemingly or actually visited you? Apparently this happened to me. Here is what happened.

I had a friend at work, a joyous professional of Hispanic decent. His name was Expedito M. Rarely does one meet someone so full of life, so joyous, so unpredictably pleasurable. Just walking with him was a delight. For example, while walking with him during a lunch break, I noticed a ladder in front of us. As I passed the ladder I found that Expedito was gone. Where was Expedito? He had scampered up the ladder and was on the roof. This was a man in his 50s. At another time, in a department store, he asked a salesman to get him a pair of shoes to try on. When they arrived he put one shoe on, the remaining shoe remaining tethered to the first by its laces. Then he got up and walked, the tethered shoe flopping every which way as he walked. Then he faced the salesman and complained that something was wrong with the shoes, they were not comfortable and acted strangely! This was Expedito: his main desire at the time was to seduce a plain looking and serious Nun who was working in the firm for the summer. His other joys were traveling abroad and his yellow jaguar sports convertible.

But he had frustrations as well. A newly hired covetous Vice President wanted to incorporate

Expedito's position as head of international sales into one of his own responsibilities, cutting Expedito out. This clearly threatened Expedito who loved his position and loved the adventure of travel. And just as clearly, he despised the covetous V.P.

But inexorably the jealous and unscrupulous V.P. started to take over his responsibilities eroding Expedito's position. And Expedito, as you can imagine was absolutely furious about this. His love of life, his pride in his professional status and his joy of travel were all consuming, in essence his position defined him. So it was not surprising that one evening after a confrontation with the V.P. he leaped into his car, his beloved yellow jaguar convertible, and took out his frustrations with speed; unfortunately too much speed. He lost control, the left side of his car hitting the barrier separating the traffic moving in the opposite direction. As a result his vehicle turned over at high speed. There were no roll bars on the vehicle and his head met the pavement splintering as the car careened to a stop. Expedito died (as we know death) that evening but I do not believe his lust for life died. For that very evening as my wife and I were driving home to Little Neck on the Cross Westchester Expressway, unaware of his death, two extremely unusual and inexplicable events started to unfold. Suddenly the car to my left lost its right front wheel rim, the kind with 3 cutouts around the edge of the rim. The vehicle behind the one losing the rim captured the spinning object under its front axle causing the rim to throw off a shower of sparks as it scraped along the pavement. The sparks backlit the three openings of the rim presenting a picture of a smiling jack-o-lantern type configuration. My wife and I were impressed by the startlingly unusual display. Coincidence? Well, within the next few minutes another vehicle lost its rim which attached to the front end of another car. The odds of this occurring at all and in such rapid sequence are just about nonexistent. Expedito joyously saying farewell to a friend? Probably.

Another unusual incident occurred years ago one late evening in Woodland Hills, California. I saw what I presumed to be a satellite, a bright spot moving stately across the night sky. I called my wife and pointed out the "satellite" which she was able to find and enjoy. Suddenly, when the object was almost directly above us, somewhat to the West, it stopped and remained stationary for several minutes. Then just as suddenly it shot off at a right angle across the heavens at an incredible speed. Our advanced technology in 1980? I doubt it. A balloon, impossible. Alien technology? Time travel? Who knows? I only know what we saw; an unidentified flying object. A normal part of our existence? Possibly, but if so, I am not aware of it. But it happened.

Then on another occasion, this last summer, as I drove down Avenida Vista Grande, a major avenue in Eldorado, Santa Fe there was a women jogging in front and away from me to the left on a walkway. The day was sunny, her path clear, there were no turnoffs and she was easily seen. Suddenly it appeared that she was starting to get smaller although my car was approaching her much faster than she was running. And this continued, she kept getting smaller, and smaller and finally She Was Gone! Disappeared! The distance was less than 200 feet from the car and she was gone. A delusion? A Mirage? Then how was it that my wife who was with me saw this happen as well?

So then, what is reality? I am a PhD Microbiologist-Immunochemist, now retired. I believe in the tenets of proven truth; Koch's postulates – find it, isolate it, grow it, transmit it, confirm it and repeat it at will. I also learned about some of natures subtleties. For example precipitating antibodies to renin, a kidney protein is produced by Dutch Belt rabbits, but not by New Zealand white rabbits. This is undeniable truth, absolute and confirmable. But what is the absolute truth about our existence? Of this we are uncer-

On the Dual Reality of Existence

tain. We have our faiths, we know that things happen, at times unexplainable things, but we are uncertain, unknowing of what is actually true reality. Is our faith true? Is virgin impregnation as an act of God, or ancient gods, myth or reality? Is reality, as we know it real? Are we truly mortal or are we immortal? The final answer is that we do not know with certainty, we may have faith, we may have beliefs, but in terms of absolute and irrefutable truth, we just do not know.

Here is a final example of an experience I had that I cannot possibly explain. When I worked for a pharmaceutical firm in New Jersey I had essentially a one-hour commute to my home in Laurel Ridge, Tuxedo Park, N.Y. On returning home I made note of a manhole cover shortly before a sporting goods store that meant to me that I was approximately ten minutes from home. I worked for this firm for over fifteen years and had had this habit of observing this manhole cover over these years; that is until about after the 12th year when an anomaly occurred. There was suddenly no manhole cover before the sporting goods store; instead there was a manhole cover AFTER the sporting goods store where there had been none before. There was no sign of any work having been done, the change literally occurred overnight. What happened? What happened? Was this imagined, an error of memory?

Possibly, who knows.

Well, someone knows or something happened that just cannot be explained by logic as we know it. It just happened, something caused it, and possibly someone knows WHY it happened. I don't, and undoubtedly I never will.

Be assured that the four incidents described are true, they actually occurred. This was reality, my reality augmented, but these occurrences are scientifically impossible to prove. So, what is reality, what is it that we experience as being out there and what actually is out there?

Conclusions

Most all of us have experienced one, or several, inexplicable happenings. This supports the fact that science does not have all the answers and that other factors, as yet unknown, or as yet unproven, exist. Clearly then, the search for the true nature of reality must and will continue.

Chapter Seven

Crop Circles

The first question to be answered is this: what do crop circles and other peripherally related phenomena have to do with the nature of death? The direct answer is that unexplained phenomena demonstrate how little we know about the true nature of our existence. (What is truly amazing about this is how far our technology has advanced in spite of this fact.)

Crop circles have been reported for centuries but modern reports initiated in earnest starting in the 1970's. In the earlier reports, mostly from England where most crop circles appeared, they consisted of simple circles or circles within circles. Most of these formations were relatively simple, although some were reasonably sophisticated. Eventually, however, the designs became more complicated such that even grossly falsifying them would essentially be impossible. An excellent review of this topic can be found in **CropCricles.net**.

Interestingly, Colin Andrews, an avid investigator of crop circles, provided evidence indicating that 20% of the authentic circles occurred within several degrees of the earth's lines of magnetic force. This he believed accounted for the swirling character of the designs.*(Crop Circles Signs of Contact, Colin Andrews with Stephen J. Spignesi, 2001; Wikipedia, crop circle, 2012.)*

Eyewitness accounts of this phenomenon in the process of forming are rare as most crop circles appear overnight. There is evidence however that a change from a normal field to a formed crop circle within that field could be essentially instantaneous. A moment ago they were not there and then suddenly they were in place.

One of the more striking examples of this occurred about 200 yards away from Stonehenge (*www.colinandrews.net/JuliaSetStory.html*). In July of 1996 on a sunny day a crop circle, purportedly in the shape of a Julia Set fractal, suddenly appeared. As the area was close to Stonehenge many people were about. Site guards had been patrolling and there had also been recent flyovers. Investigation indicated that at most the formation had appeared some time within a 15 minute interval. The complexity of the formation would have taken engineers weeks to duplicate the overall design as closely as possible. In this case, as in most cases of authentic crop circle formations, it is extremely unlikely that the detailed nature of their unusual features could have been fraudulently duplicated.

Most other reports, by the thousands, have not been as clearly witnessed, some reported the circles occurring in the presence of bright spheres in the sky emitting rays of light, others reported a buzzing sound and still others claimed to have experienced a sensation of being charged as with static electricity.

Over the years, designs have become much larger and more complex (why this should have occurred remains unexplained). Many of the patterns formed are now so complex that one could only assume that advanced mathematics was used in the design of their formations.

Crop circle formations have now been reported from all over the world. Some indeed had been made by people, for fun, to confuse, or to obfuscate the existence of true crop designs. Some believe that deliberately manufactured fraudulent crop circles were being produced by government agencies to cover up the fact that they are real. If this were true, this claimed government cover up implies hidden significance to this phenomenon, perhaps hinting at significant extra-terrestrial activity and communications.

Frankly, I do not accept this conspiratorial explanation. As far as I am concerned, claims of government interference to discredit the existence of this phenomenon are unfounded. Nonetheless I also believe that crop circles, authentic crop circles, are real and truly exist. I believe this because clear, indisputable and reproducible differences exist between man made and real crop circles. I believe this because evidence exists that they may result from natural phenomena.

For starters, the geometry of authentic crop circles is perfect. No false copies can duplicate this perfection. Another key fact is that affected plant stems are bent, not broken as is found in man made designs. As stems are bent and not broken, plants are not killed. In falsified crop circles made by people, affected stems are broken by planks laid upon them used to form the design. Here, affected plants are severely damaged or killed.

It has been noted that in authentic crop circles

stems seem to have been heated at about one inch above ground and often vacuoles are formed at or within the stem at about this point. Also, soil adjacent to the formations is found to be richer in iron than soil further away from formations.

Photomicrographs of affected plant stems and their surroundings show microscopic rounded beads of iron, at times imbedding in the bent region of the plants, or in their surrounding, immediate vicinity. Also very significant, and what I believe to be a critical associated phenomenon, is the presence of magnetized iron spheres in some stems and also in and about locally fused soil. Occasional findings of short-lived radioactive activity have also been noted. It is important to note here that it is well known that meteorites are rich in iron. In fact, meteorite hunters use metal detectors to find them.

It is also a known fact that meteorites of varying size and composition are constantly bombarding Earth's atmosphere, but not all meteorites hit the earth. Most are small and are vaporized; however, some (fortunately rarely) do hit the earth and are large enough to cause catastrophic damage. Clearly there is a range of meteorite sizes from extremely small to very large all of which eventually impact Earth. Is it not then very probable that some hit the earth as microscopic particles? Magnetized particles *(see gravityc-idealism.blogspot.com/2007/02/crop-circle-dynamics.html)*?

When a meteorite traveling extremely fast hits our atmosphere it becomes molten hot due to friction, hot enough to vaporize it in part or in its entirety. Then these solid and vaporized particles hurtle toward Earth due to the innate kinetic energy they possess and the effects of gravity. Some arrive as solid visible meteorites, some dissipate and some I believe hit the Earth as microscopic particles and with significant heat energy. It is quite probable that disintegrating iron particles cool to a degree and form microscopic spherules when passing through atmospheric water vapor and clouds prior to hitting Earth. The magnetic nature of these particles can be visualized as occurring due to friction in their passage to earth. This would be similar to what can happen to a glass rod by rubbing it on a cloth; it can act like a magnet due to static electricity.

One can then reasonably assume that distinct microscopic packets of almost vaporized matter and the heat energy developed in them are the basis for crop circles.

Well then, if all the above is what actually occurs, why is our planet not covered with crop circles? Reason suggests that crop circle formations are relatively rare phenomena requiring conditions that are just right for them to manifest. Goldilocks conditions: not too large, not to small, and atmospherics just right. This however does not explain the varying shapes, sizes and complexity of authenticated circles. So can this be explained and if so how?

Picture this, an iron bearing meteorite, composed of different minerals of just the right size, hit the atmosphere and send out a shock wave and particles the initial shape of which is dependent on the angle of entry. The meteorite heats up and starts to fall apart. As these fragments fall through the atmosphere those containing iron particles become magnetized by friction with existing atmospheric dust; they ball up into microspheres. Still extremely hot the particles fall through the atmosphere at supersonic speed while now being cooled to a degree by water vapor. As some velocity is lost the overall shape of the descending particles is modified by winds, by varied resistances in the atmosphere at that time, by magnetism induced grouping of particles and by their proximity to the lines of the earth's magnetic field. In seconds, the entire mass hits a field in its patterned shape, the remaining heat quickly dissipating but sufficient to slightly scorch and expand the moisture in plant stems (causing vacuoles and causing them to bend) the composite affect thus imprinting a unique pattern on the ground. Voila, a unique crop circle is formed. And indeed, aspects of this hypothesis are starting to be discussed on the internet.

Now let us consider the snowflake. Snowflakes exist in myriad shapes and forms depending upon atmospheric conditions including temperature, moisture content, clouds, turbulence and innate energy.

An entering vaporizing meteorite packet meets the same conditions. However, as these iron rich particles cool to form microscopic spheres they also are becoming magnetized due to friction generated by their passage through the atmosphere. Thus we have this descending shower of particles whose configuration is dependent upon all the forces acting on them, much as what happens to snowflakes. So, let's consider snowflakes as well as crop circles

Are there similarities and differences? In meteorites, as against what we get in snowflakes, there is added complexity. Here we have to account for magnetism, differences in the angle of entry, the relative proximity to earth's magnetic field lines, fracturing, heat and kinetic energy; all these factors additionally affect any crop circles' eventual form and distribution. Clearly, these effects could easily account for many of the unusual crop circle shapes. Interestingly though, whereas crop circles form designs that can repeat themselves, the shape of every snowflake is different. The significant point is that snowflakes are accepted natural phenomena and that in all probability it is likely that crop circle formations result from natural forces as well.

It must be admitted, however, that some designs are so complex that it is difficult to imagine that natural forces formed them.

Although I do not pretend to know the actual mechanism involved in crop circle formation in every instance, I believe that the mechanism described here provides a logical and reasonable hypothesis for many of the authentic formations.

In summary then, not all crop circles are false and man made, not by a long shot. And most probably they are not formed by intelligent extraterrestrials sending us coded messages; nor is it likely that they are signs of life from Gia, the living Earth.

Conclusions

If the above hypothesis is correct, it indicates that many crop circles are not the result of paranormal happenings and therefore, they shed no light on metaphysical phenomena. They are simply rare and beautiful natural occurrences.

The existence of exceptionally complex formations, however, does remain unexplained and their implications are unknown. If they are authentic, it may be possible that they indicate an unknown mode of action and therefore extra-terrestrial activity cannot be totally ruled out.

Chapter Eight

Animal Mutilations

Beginning in the 1967 horrendous occurrences of killing and then mutilating cattle and other livestock began in Pennsylvania, central and western United States and even earlier in England. *(www.noufors.com/animal_mutilations_a_worldwide_phenomena.html; http://vault.fbi.gov/Animal%20Mutilation)* Sheep and horses have also been found dead and mutilated. There is even a reported case of one human corpse having been mutilated in a similar fashion. Although instances of their occurrence have diminished, mutilated animals have now been found world wide, in South America, Europe and even as far off as Indonesia.

The nature of these mutilations has been extremely unusual and bizarre. In most cases all blood had been drained from the animals and their sexual organs, left ears, left eyes and intestines have been removed. It is not known how these animals were killed but all tissues that had been removed were done so with surgical precision. Also baffling is the fact that no footprints, tracks, or any other indication of the interdiction of any animal or mechanical devices surrounded the bodies, even in muddy soils (although strange tripod like marking were sometimes found near the bodies).

The wounds, made with great precision appeared to be cut with exactitude by a trained intelligence using surgical instruments. No blood appeared at the margin of these wounds where the cut lines were clear, regular and appeared to have been cauterized. As previously mentioned, in most cases no blood was found anywhere.

There have, however, been some differences found where (more, or less), specific tissues were taken from individual animals; some missing their tongues, their lips, their hearts and other varied internal organs. All tissues that had been removed had clearly been done surgically.

Local authorities investigating these occurrences

were baffled by the lack of signs that predators had inflicted these wounds. There were no tooth marks, no rendered torn flesh and no animal footprints. In fact, predators, livestock and pets avoided the corpses for days. They were apparently frightened by something about them. Also of interest is the fact that some of the carcasses appeared to have been dumped from a height causing a slight depression in the soil under their bodies. This indicated that the bodies had been airborne prior to being dropped and perhaps had been mutilated elsewhere, thereby accounting for the lack of any blood.

The cause of these mutilations and associated phenomena remains a mystery. Conclusions from a series of private and government investigations vary widely, from the mutilations being due to natural causes, to predators, to unstable individuals, possibly to cults, perhaps even to secret government actions and even to extraterrestrials. There is no agreement. Claims have been made that strange aircraft have been seen at times in formation, flying at high speed and generating humming noises. Some claim they have seen flying saucers near bodies; others reported the appearance of black helicopters, indicating government activity. (Clearly, there is no need for a government agency to clandestinely kill thousands of animals.

If experiments of this nature were required they could be funded and carried out at government facilities).

What I believe is that in all probability these mutilations may have multiple causes, I also believe, that many of the classic cases have not been explained and therefore are of unknown origin. (More detailed information in regard to these occurrences and investigations of them can be found on the web by googling "animal mutilations").

Conclusions

No absolute conclusions are possible at this time, thus indicating just how little we know about animal mutilations as well as other unusual happenings. The possibility that alien "anthropologists" seeking to learn the nature of life on Earth are the source of these occurrences must be considered. Undoubtedly, alien intelligences would want to know the "workings" of newly found life anatomically, metabolically and genetically. There is apparently no relevance to spirituality attached to this phenomenon.

Chapter Nine

Unidentified Flying Objects

Referring to UFOs, the primary question revolves about the nature of reported sightings. Are they real or do they reflect misinterpretation of natural phenomena? This question is additionally complicated by the fact that reported sightings of UFOs have literally spanned history.

Nonetheless, what is clear and without question is that a large number of reports, probably most of them, are based on poor observation. Most can and have been explained away by proving that the actual nature of these sightings were misinterpretations of natural phenomena.

However, there remain a number of reports that

are unexplained. And many of these unexplained reports were confirmed by radar, reported by experienced commercial pilots, cited by skilled air force pilots in scrambled jets and even confirmed by astronauts (*http://en.wikipedia.org/wiki/Unidentified_flying_object; http://en.Wikipedia.org/wiki/Roswell_UFO_incdent*).

These unexplained reports have been listed as being "unknown" by official (Blue Book) authorities headed up by Edward J. Ruppelt. It was Mr. Ruppelt who changed the name of such sightings from being called flying saucers, which implied extraterrestrial origins, to UFOs, which simply stated that the reported sightings were unidentified. This direct step avoided the emotional issue of possible extraterrestrial involvement.

When last reported, the opinion of Mr. Ruppelt, was that although there have been many unexplained UFO sightings they probably resulted from the misinterpretation of natural phenomena. He also pointed out that as far as he was aware, there had been no official hard evidence proving the existence of UFO's *(www.nicap.org/rufo/contents2.htm)* – although many unconfirmed claims existed. It is well known, for example, that several of our presidents, such as Ronald Reagen and Jimmy Carter claim to have seen UFO's,

On the Dual Reality of Existence

they have also been seen in other nations and again, even some of our astronauts have officially reported seeing UFOs while they were in orbit.

Mr. Ruppelt's conclusions were reached in spite of these facts and in spite of reports of noted sharp maneuvers of the UFO's, reported sudden acceleration of these crafts to avoid pursuers, (including astoundingly slow or unbelievably fast flights, and incredible high altitude flights). Reputable and professional individuals, as noted, made these reports. Thus, Mr. Ruppelt's opinion runs contrary to the observations of many groups of people, radar identification and fighter pilot confirmation. Clearly Mr. Ruppelt is jaded in his opinion, this undoubtedly resulting from the large number of mostly incorrect reports he has investigated.

So then why was he convinced that the sightings were all of misinterpreted natural phenomena? The clincher for his opinion, which he pointedly made, resulted from the fact that UFO's have never been reported by the numerous circling satellites as having entered or left the atmosphere.

But is he correct? Possibly, but his opinion is solely based on our currently known technology. Advanced civilizations may have mastered time travel or have learned to travel through wormholes, or have simply developed an invisibility cloak. Our satellites would not note crafts capable of using any of these methods.

There are however, other problems relevant to the existence of UFOs such as having demonstrated sudden incredible acceleration, severe turns, or both of these phenomena. These maneuvers would not be survivable by humans or conceivably by any living organism. Humans experiencing these affects would undoubtedly be killed. Clearly then, if UFO's exist, they are the result of visitations by extraterrestrials from an advanced civilization using technology currently unavailable to us. How then could they produce crafts capable of traveling incredible distances and able to perform, what are to us impossible maneuvers?

If we consider the possibility and capabilities of advanced technology there could be several methods available to protect the lives of those in the craft. Surviving these violent changes in acceleration or direction could be accounted for by a vessel capable of causing time shifts. This would give the craft the appearance of impossibly severe maneuvers. Or the craft may have an antigravity device, an internal force field, an anti-inertial system or composites of such systems.

No, actual sightings of real UFO's have not been ruled out. Far from it, there have been too many reliable reports of both individual and mass sightings. Radar and visual reports have even been reported over Washington D.C.

I believe that it is reasonable to assume that the capabilities demonstrated by these craft are so advanced that they do not represent secret military aircraft. I further believe that they are of extraterrestrial origin. And if this is the case, what are they here for? Are they exploratory in nature studying an "alien" civilization, are they monitoring or modifying us to influence and hopefully advance our development? Are we being visited by vessels from more than one civilization? Are they all benevolent? After all there are more than one hundred billion stars in our galaxy and over one hundred billion galaxies in our universe. Staggering! And then there is Roswell. Who do you choose to believe? All agree that an aerial device crashed in an arroyo north of Roswell on July 4th, 1947. The government occupied the crash site on the 5th, blocked any unauthorized visits to the site, and followed this up by intimidating early witnesses and then by sequentially providing conflicting and at times absurd reports as "evidence." To date freedom of information records requested and released on this event have been a maze of blacked out sentences and paragraphs.

But why the secrecy and intimidation if what had been found at the site was a weather balloon, or a high altitude test with dummies, or a craft designed to crash in the USA with deformed children from the Soviet Union sent to confuse, confound and cause panic?

All absurd, and all unfortunately the typical simplistic thinking of midlevel deskbound military bureaucrats.

Contrary to the inconsistent government reports, however, you may choose to believe the at least 6 different witnesses, (some unaware of each other), who saw the crash site and bodies. All described frail, smallish alien like creatures with large heads and eyes, a slit for a mouth, a flat just about nonexistent nose with openings for ears, no earlobes and longish fingers. Other witnesses, also unknown to each other, described metal debris with unusual properties at the crash site and saw the alien bodies. One claims to have participated in autopsy procedures. Two other reports claimed to have seen a living "alien." (It is interesting to note the diminutive and relatively frail nature of the aliens. This would indicate that this species had spent a majority of their time in space or have come from a smaller or less dense planet with less gravity.)

Weighing the evidence presented by both groups I tend to believe the eyewitnesses. What is of interest here is that the external features of these beings are somewhat similar in form and shape to that of humans. (Although unconfirmed, the sketchy autopsy results "reported" seem to indicate that the internal construct of these creatures is very different from that of humans).

Clearly, the preponderance of evidence indicates that the craft and its occupants are from an unearthly source. However, time travel cannot be ruled out.

Reports of abductions, examinations, insemination and lost time are difficult to verify. This, however, might be expected to be utilized by alien visitors, or future time travelers, to study and evaluate different life forms or to fortify their frail bodies by creating hybrids with human DNA and multifunctional stem cells.

Conclusions

The majority of observations noted by responsible trained individuals and groups confirm the existence of UFO's that are advanced aircraft of unknown origin. They are real, they do exist and they are controlled by living beings that apparently are not human. Therefore they are controlled by aliens, or time travelers, or both. Their existence provides clear and certain evidence that at least one civilization exists that is at the very least technologically far superior to ours. Quite possibly they are responsible for animal mutilations since intelligent alien beings would undoubtedly wish to learn as much as possible about other life forms.

Are the aliens messengers from a world of spiritual existence; very unlikely. From apparent evidence spiritually (gestalt) based entities can exist anywhere at any time, they would not require intermediate messengers nor would they require spacecraft. Do the aliens provide evidence of life after mortal death? Have they achieved immortality? There is no evidence supporting these possibilities. Apparently they are mortal as reports indicated that all but one died in the crash.

Aliens may not be immortal, but now that we have accepted their reality and observed their advanced capabilities it is clear to us that Earth bound civilizations have much further to go. However, our technology is advancing at a staggering pace.

As a result of our increasing knowledge and recognition of the complexity and possibilities present in the universe we are a giant step closer to evaluating the nature of life, the reality of death and the possible presence of an immortal spiritual existence.

Chapter Ten

Remote Viewing

Remote viewing is a phenomenon where individuals, almost always under controlled conditions in laboratory studies, can visualize and describe items or locations unknown to them on request *(http://en.Wikipedia.org/wiki/Remote_viewing)*.

For example, trained viewers may be asked to describe a location where they have never previously been, asked to read contents of a sealed envelope, or asked to find the location of missing persons. The point being that the viewer can literally see (sense?) and describe a scene, smell its odors, hear its sounds and taste any existing edibles present there. Apparently, time is not a barrier so that past events can be visualized such as the circumstances of a crime. Their determinations have been given orally, presented in writing, demonstrated by drawing sketches of locations, or by all of these means.

These claims, however, are by no means unchallenged and presented evidence both pro and con exists. Large numbers of studies have been performed to study this phenomenon. Carefully controlled studies have been found that apparently negated earlier reported sessions. In these studies it was claimed that in the earlier studies subliminal clues had been inadvertently provided to the remote viewers being tested. These latter investigators claimed that the clues provided to the subjects increased their accuracy and when these clues were removed the subjects' results were negative.

Reportedly most studies have now stopped; however, some continuing investigations claim positive results under very strictly controlled conditions. Details in regard to these studies can be found on the internet (for example Wikipedia, the Free Encyclopedia). Suffice it to say that there have been some very accurate readings which remain unexplained.

Conclusions

Remote viewing, when and if accepted as being confirmed, would provide key evidence of extra-corporal perception. This could be an interesting step toward believing that there is a spiritual aspect to our existence. However, this would not be conclusive as not enough is known about the physical brains' ability to zero in on information from the 'ether'. By way of example, remote viewing could be explained as being natures' equivalent to computer cloud technology (for computers, storage in remote servers).

Essentially, the simple fact is that remote viewings' existence is at best, controversial. And aside from this, it cannot and does not clearly provide any firm evidence of a spiritual existence distinct and separate from animate existence.

Chapter Eleven

Out of Body Experiences

In essence, an out of body experience is the sensation of leaving ones body while maintaining clear visual and auditory ability. Most report having left their bodies, followed by a sensation of floating upwards toward the ceiling. They then can see their separated bodies underneath them. Out of body experiences (OBEs) are not like dreams. They seem very real and are clearly recalled for many years.

    Dealing with a topic as intense as this, as you might imagine, has been subject to a great deal of controversy. Studies and evidence both supporting OBEs as being real as well as refuting their occurrence as being hallucinatory both exist. As is usual in this text I will present only the most compelling evidence. And a most compelling example in favor of the reality of OBEs is the case of Pamela Reynolds *(http://en.wikipedia.org/wiki/Pam_Reynolds_case; Light and Death ISBN 0-810-21999-2* by Dr. Michael Sabom).

Ms. Reynolds required brain surgery as the result of having a massive basilar artery aneurysm. This is an extremely difficult and dangerous procedure as the aneurysm can be easily ruptured and lead to death. Dr. Spetzler, an expert neurosurgeon, was selected to perform this procedure as he had perfected a method called "hypothermic cardiac arrest" that could repair severe aneurysms in dangerous locations with much less risk. The procedure he developed is extremely complex and details will not be presented here. The method involves cooling a patient to the point of clinical death and then draining their blood to collapse their aneurysm. At this point the patient's aneurysm is removed, their blood vessel repaired, their blood slowly warmed and returned to their body. Pamela underwent this procedure. During this 35-minute interval she had her OBE and a following near death experience.

Pamela described her experience as a series of events, first 'hearing' sounds apparently associated with the surgeon drilling through her skull. Just after that sensation her OBE initiated and she felt that she had left her body and was looking down at the procedure from above. It is most interesting to note attempts to ascribe her experience to natural causes. Presumptive reasons were given to explain her ability to "hear" drilling sounds and her ability to accurately describe a conversation between her nurse and surgeon (all while under anesthesia, with both ears occluded and with acoustic plugs producing periodic pinging sounds to assess her brain function on an attached encephalogram). Failure to respond to the pinging sound would indicate cessation of conscious brain function.

All of the skeptical reasons given are reasonable and claim to be scientifically based – but all of these de-mystifying explanations fail to explain one thing. How was Pamela Evans able to visualize the surgeon drilling into her skull during the procedure?! How Pamela Reynolds was able to, in essence, properly visualize and later describe the instrument the surgeon used to cut out a section of her skull? How was it possible that she could so accurately describe an instrument she had never seen before? How could she have so accurately described this instrument when she did not even know that an instrument of this type and configuration existed? How could she do this when unconscious? Did this visualization occur when her body temperature was still normal and when her heart was still functioning normally and with an intact circulation?

Did she hear sounds probably caused by the surgeon drilling into her skull? Yes, this was quite possible. Her ears may have been plugged, but even under anesthesia bone conduction with the sensation of sound was possible and although anesthetized, her other body functions were normal. Consciously hearing and accurately describing the nurses' conversation while under anesthesia, with ears plugged and a device making beeping sounds? Although this is unlikely it may have been possible if the anesthesia she was given was somewhat light and if the earplugs had been improperly placed. But visualizing and being almost entirely accurate in describing the instrument being used by the surgeon when she was unconscious on the operating table, IMPOSSIBLE unless she had watched the surgeon from above during an out of body experience.

Skeptic's attempts to provide possible materialistic reasons for why she could describe the saw (from having seen dental drilling tools) are weak to the point of desperation.

Following her OBE, Pamela had a near death experience that we will discuss that later in this text. The main contention here was whether she was simply unconscious or clinically dead at the time of this experience. Unfortunately, conflicting assumptions have been made that cannot be verified one way or the other.

As a result of Pamela's' reported experiences, attempts were made to test the reality of OBEs during other surgical procedures or under laboratory conditions, but these have consistently failed. These tests consisted of placing numbers or photos on the topside of plaques suspended near the ceiling. In every case save one, those claiming to have had OBEs failed to identify the photos or the numbers.

In an experiment conducted by Dr. Charles T. Tart a single positive result was obtained where the possibility of a confirmed OBE existed. In this case the subject correctly identified the numbers as 25,132 that were written on a slip of paper placed face up on a high shelf that could not be seen by anyone standing in the room. On careful investigation however it could not be ruled out that the subject may have been able to see a reflection of the numbers on the black plastic case of a clock mounted on the wall. Although this was unlikely, it could not be ruled out. (Dr. Tart has published 2 classic texts in this area that you may wish to read: ***Altered State of Consciousness (1969) and Transpersonal Psychologies (1975))***.

Hard, verifiable, evidence is difficult to come by, but does exist the case of Pamela Reynolds being a prime example of this fact.

Then there is the case of Maria who suffered a heart attack and cardiac arrest. She told her social worker, Kimberly Clark Sharp ***(www.spiritdaily.net/clark.htm)*** that she had an OBE while unconscious and from the ceiling watched the cardiac resuscitation teamwork on her. She also said that she found herself outside of the hospital where on its north side she noted a tennis shoe on a ledge on the third floor. She then saw that the shoes laces were undone and that one of the shoes laces was underneath its heel and that shoe showed wear by its small toe area.

Although skeptical, at Maria's request Ms. Sharp looked for, found, and retrieved the shoe confirming Maria's vision. She further stated that Maria's detailed description could only have been seen if she were floating outside the window and seen the shoe at close range ***(After The Light William Morrow & Company, Inc. 1995 ISBN 0-595-28028-5)***. Other clearly veridical evidence has been presented by Dr.

Raymond A. Moody Jr., who has written a series of texts on the subjects of OBEs and NDEs (See, for example, his text (*Life after Life ISBN 0553274848 (0-553-27484-8)*. In these texts he describes a number of these experiences, read them if you wish to learn more about his many findings of verifiable OBEs.

Well, at this point I believe that we have spent enough time on this topic; hopefully, more than enough to confirm the reality of OBEs.

Conclusions

There have been approximately 25 million reports of OBEs in the last 50 years. Still, this topic remains controversial, undoubtedly because laboratory or planned studies have given negative or at best possibly positive results. Also, not all spontaneously reported OBEs are real or accurately reported. Some occurrences take place while individuals are dreaming or during crises, others are dream state constructs or are hallucinatory. This confounds confirmation of the phenomenon. On the other hand, numerous spontaneous reports of OBEs have been verified.

I cannot explain the discrepancies found between formal studies and spontaneous reports, nor can I totally dismiss the existence of imagined or unsubstantiated OBEs, ASTRAL Projections (consciously directed

OBEs) or results obtained through claims that such experiences can be learned and used on demand. Nonetheless, the large number of verifiable results cannot be overlooked.

Therefore it is concluded from the verified evidence obtained that OBEs are a real phenomenon and indicate the presence of a dual aspect to life, material and spiritual.

Chapter Twelve

Reincarnation

Here we are starting to approach the heart of the matter. Reincarnated individuals by definition have had a spiritual transfer of memories from a past life into their current life. If this existed it would appear to definitely confirm, at least the existence of a temporary, spiritual existence. IT HAS BEEN CONFIRMED, read on.

One must then ask, when information is transferred from a deceased individual to a living person is this evidence of a spiritual existence or is this as yet a not understood natural manifestation of a cloud-like phenomenon? Frankly, the latter hypothesis seems unlikely because of the detailed knowledge the recipient has covering years of experience of a past life's' existence.

Early records indicate that the concept of reincarnation originated around 1200 BCE. Since then and over the years in slightly modified form, the concept could be found in Egypt, Tibet, India, Greece, by Celtic Druids, in early aspects of Christianity, in Judaism, in certain Shiite Moslem sects, in Buddhism and in current belief systems such as Scientology.

Some of those claiming to have been reincarnated said that they were reborn in bodies of the opposite sex, or in a different religion or in a different nationality. The significance of this will be discussed later. Current (and what I believe to be uninformed) thinking in large part supports the belief that evidence presented in support of reincarnation has been unreliable, essentially anecdotal and basically unscientific.

So then what is the truth? A psychiatrist from the University of Virginia, Dr. Ian Stevenson studied reported instances of reincarnation in about 2,500 young children from around the world over a 40-year period (from the 1940's thru the turn of this century). He compared the children's' stories of past lives they claimed to have had with factual information about those deceased individuals. Amazingly, he found that in most cases the information provided by the children

On the Dual Reality of Existence

factually matched events in the deceased individuals lives. Direct contact or prior knowledge by these children of the deceased personages was demonstrated to not have been possible in most instances. One of his basic texts on this topic is *(Twenty Cases Suggestive of Reincarnation University of Virginia Press 2nd ed. 1974)*.

As might be expected his efforts were criticized as being "pseudoscientific" and unreliable. Frankly however it is difficult to believe that information of this type could not be valid or that it lacks scientific integrity. It must be recognized that any attempt to apply absolutely rigid scientific methodology in studies of this type would be extremely trying, if not totally impossible.

Many claims of reincarnation have been made that have been confounded by preconceived religious teachings or essentially have been put forth with inadequate and insubstantial evidence. And this has provided ammunition for skeptics who will by their very nature always attempt to discredit individuals and concepts they strongly disagree with. ("I do not believe it so it cannot be true," hey there, that's not a very considered or scientifically based conclusion).

The skeptics, however, have a problem accounting for a recently reported instance of reincarnation

On the Dual Reality of Existence

(FROM SOUL SURVIVOR by Andrea Leininger, Bruce Leininger, Ken Gross. Copyright © 2009 by Bruce Leininger and Andrea Leininger. By permission of Grand Central Publishing. All rights reserved). You may well be familiar with the story.

James Huston Jr. a WWII navy pilot could have been forgotten, but for James Leininger. Starting at age 2, James began to have nightmares exclaiming *"Airplane Crash! Plane on fire! Little man can't get out."* James told his parents that in his past life he was a pilot, also named James, on the carrier Natoma and had been shot down by Japanese antiaircraft fire. Investigating James' story, Mr. Leininger learned that a pilot on the Natoma, James Huston Jr., had been shot down in 1945.

James mother located James Huston Jr'.s sister, who later spoke with young James. During this conversation he presented detailed information relating to James Huston Jr's life. He also asked about one of two matching portraits of his "sister" and him done by their mother.

There is no way he could have known of the existence of these portraits unless he possessed the reincarnated spirit of James Huston Jr.!!!

All this information and many more significant validating details can be found in the text written by the Leiningers and Ken Gross.

On the Dual Reality of Existence

As you might imagine, sharp criticism and efforts to debunk this story have been made. The first assumption made by critics is that the premise of reincarnation has no basis of fact. Another is that the child recalled snatches of information garnered from an early visit to an aircraft museum (at the age of 18 months) and the third is that he constructed this fantasy on being prompted by a spiritual counselor.

As part of any investigation, the possibility of fraud or innocent leading questions conducive to the assumption of a reincarnated spirit must be considered. But in this case they can be ruled out because of his request for the portraits.

Suffice it to say that attempts at debunking this story are based on assumptions and premises that are as incredulous, or even more so, than the reincarnation story. The claim that the name Natoma arose from prior knowledge (by a 3 year old?) is bizarre. How did the child know that James Huston Jr.'s aircraft was hit by antiaircraft artillery? Was this by "coincidence?" The fact that others who flew the same mission confirmed this story is also unexplained by debunkers.

To John Huston Jr's sisters' amazement the child came up with facts about her brothers' life that should have been unknowable. True, many of the family facts could have been retrieved from archives, but again, how could he have known of the two matching portraits of his "sister" and himself (James Huston Jr.) done by their mother?

Another case, for example, was that of a young British child claiming to have been a world War II German bomber pilot who was shot down over London, crashed through a window, and was killed. *(http://www.iisis.net/index.php?page=semkiw-ian-stevenson-reincarnation-past-life-carl-edon&hl=en_US)*. He proved to be very familiar with the instrument panels of German aircraft and with many other things German. Years later the body of a dead German bomber pilot matching the description provided by the child was found in London.

Although this case is not as strong as that presented by the Leiningers, it is quite interesting, in that it supports the concept that one can be reincarnated in a different nationality (and possibly also in a different race, religion or sex).

And there are of course many thousands of other instances of claimed reincarnation, but none I believe more compelling than the story of young James.

Conclusions

The presented evidence strongly supports the existence of reincarnation no matter how unrealistic and improbable this phenomenon may seem. Undeniable fact is just that, undeniable.

Reincarnation happens.

Confirmable evidence supports spiritual survival after death. Therefore, it exists and is real. And reincarnation could not be possible if there were not a continuing entity coexisting in a living carnate body that separates when corporal existence ends in death.

Chapter Thirteen

Near Death Experiences

Near death experiences are real. It is their cause that is under question (*www.nderf.org*). Is this a manifestation of a dying brain or is this ones spirit leaving its body? In this matter interesting proofs do exist although a number of formal skeptical studies indicate that some NDE-like phenomena can be reproduced in laboratory settings. But then, how does one evaluate Steve Jobs last words; "Oh Wow, Oh Wow" are these the utterances of a dying brain?

Consider the fact that we already know spiritual beings can depart from their bodies. This has been documented with confirmed evidence both in OBEs and in reincarnation. This proof has been established.

On the Dual Reality of Existence

Undeniably, there is spiritual existence after death. But although the validity of these phenomena is undeniable we have not as yet proven the nonmaterial nature of NDEs. So how can we evaluate and prove that NDEs are real and that the spiritual visions seen of deceased loved ones in a utopian existence are real and not hallucinatory?

Most of us are aware of the nature of near death experiences and deathbed experiences. They occur when someone is apparently dying, or is reported as being clinically dead. Of course they are not truly dead as some have not yet died and others can be restored to life.

On the other hand, when you are dead, you are dead. Unfortunately, and by definition, when you are dead there is no coming back to confirm the occurrence of an NDE.

Well then let's look at the evidence such as it is and define those reporting having had a death bed experience or an NDE as the 'almost deceased'. NDEs and deathbed experiences occur in every race, ethnic group and religion in the world but although many people have been close to death, only about 23% of those (about 25 million) have had such experiences. Does this mean that those who have not had them will not survive death as a spirit because they

On the Dual Reality of Existence

have no soul or does this mean that they were not "dead enough" yet to have had these experiences? The answer to this is not known.

Extremely interesting is the fact that NDEs frequently vary with the culture individuals have lived in. Most individuals of European descent first report a feeling of separation from their bodies followed by being in absolute blackness. This is then followed by seeing a distant light at the end of what seems to be a black tunnel and the sensation of effortlessly traveling towards the light, which grows larger as it is approached. In these circumstances they conceive of themselves as being physically intact although in fact their unconscious bodies are elsewhere, most usually in a hospital, so at this point they are spirit, or if you wish, soul. Soon the light they have been approaching encompasses everything and the individual experiences incredible peace and joy. Detailed life reviews have been reported. Finally, deceased relatives are seen as spiritual entities looking alive and well, even younger then they looked when they died. These include grandparents; other relatives the individual had never seen <u>and deceased relatives that they did not know had died</u> (but on return to their bodies they learned that these persons had indeed very recently died). Some have religious awakenings not necessarily in the beliefs they held when consciously alive and some few have terrifying experiences. The great majority however experience darkness, the light and seeing their deceased loved ones in spirit form.

The question is, are we dealing with actual spirits or with the hallucinations of a compromised brain? If these beings are real, can they assume any shape they wish? Are they indeed the spirits of our departed family, or a family of advanced beings? Is it not interesting that the sensations experienced by those having a near death experience are most usually joyful and that they wish to remain in the spirit realm? This, since they sense what they perceive to be a utopian existence where everything is absolutely perfect and joyful. If these experiences are true, and souls survive death, is a dying mortal in actuality immortal?

As stated, most individuals in each ethnic group visualize scenes appropriate for their culture, although this is not always the case. Pamela Reynolds, who's out of body experience was discussed earlier, noted total darkness, a tunnel with a light at the end and seeing her deceased relatives. Essentially, her experience is typical for Europeans and Americans.

American Indians usually experience seeing beautiful open pastoral vistas, while many, but not all, Asians from India felt that they were being guided to

On the Dual Reality of Existence

a desk by guardians of the dead and being adjudicated by a judge reviewing their life's deeds.

I believe that variation in the nature of NDEs as a function of ethnicity or expectation is required to soften the shock of knowing that "you" have died. There are very good reasons why to expect any transfer from material reality to spiritual reality to be tempered. Facing reality, dying is not easy. All that you have been and are, as far as you are concerned, will be completely obliterated. On the death of a human coexisting with a spiritual entity, a mechanism must be in place to instill a sense of joy and peace in the released spirit. This then is enhanced by the visualization of past loved ones as well as by apparent fulfillment of their heavenly expectations. Thus, any potential emotional shock caused by their transition from a material existence to a spiritual realm is ameliorated,

Once newly arrived spirits have been fully integrated into their new and more permanent existence a more complete and accurate presentation of their spiritual existence will undoubtedly be presented. It will not be negative, but it will be substantially different from their initial encounter. No physical body, no need to eat, no digestive tract, no sex, in fact no carnal sensations whatsoever, not even a sense of rain, wind, solid matter, or physical touching. But a sense of peace, eternal existence, "soular" intimacy, beautiful surroundings, growing enlightenment and a true sense of total communion.

So, first contact requires that most, if not all, in each group have NDEs as a function of their ethnicity or expectations. As stated, I believe that the shock of transition requires it. The key point here, however, is the substantial proof of the reality of NDEs. This established by the fact that some of those having had near death experiences or death bed experiences have spiritually seen deceased persons that they, or even their living relatives, did not know had died. On 'return' or on inquiry both they, if possible, and their families later learned of and confirmed the passing of the relative visualized in the spirit realm *(see for example 'One Last Hug Before I go: the Mystery and Meaning of Death Bed Visions' by Carla Wills-Brandon)*. This clearly supports and is significant evidence favoring the reality of these spiritual phenomena. In fact there has now accrued a large body of information supporting the reality of NDEs; see for example *(http://eteponge.blogspot.com/.../near-death-experiences-out-of-body.html)* for an in depth evaluation of this phenomenon or find many other confirming articles on the Internet.

Conclusions

Most, if not all, near death experiences are real events. To reiterate, this has been verified. There have been NDE cases where deceased relatives or friends that were not even known by their living relatives to have recently died were seen in the spirit realm or were visualized as being in a hospital room by dying individuals.

The undeniable conclusion, based on the confirmed evidence presented here and in many other studies on OBEs, NDEs and reincarnation is that a dual reality of spiritual and corporal existence is real. This point has been unequivocally demonstrated. Theoretically we could stop our study at this point, but that would leave many questions unanswered. How did this come to be? Is there a reason why lives are structured in this fashion?

Here there are no factual answers, only conjectures. The remainder of this text will address these questions.

Chapter Fourteen

Possibilities

There are several possibilities to be considered in determining what the dual reality of our existence is, how it came to be and why it occurred. One can consider that a supreme being created this mode of existence, or that a chance quantum variation in our DNA or cerebral structure brought it about, or finally, that it could be the fruit of an advanced technology.

A Supreme Being

The difficulties involved in considering a supreme being as the creator of our existence have been noted earlier. The Atlas problem, the different and conflicting religions, why the last significant proponents who are believed to have presented evidence of such a being lived 1500 to 3000 years ago? Why such a being would permit genocide, pain, pestilence, war, crime, hunger and cruelty?

Religious claims that we have "free will" or are being tested have a shallow, hollow, ring. And finally, there is the unfortunate lack of confirmable evidence as well as the clear mandate of the existing universe as we know it. All this pointing to there being no supreme entity. Nonetheless, this possibility cannot be ruled out. Why, because there is no answer to this question. The only truth at this point is to conceive of a flea attempting to define an elephant. In no way can this realistically be done. Here we must live without evidence except possibly for flowers, air and water.

Random Cerebral Modification

Random cerebral modifications could possibly account for the existence of souls, but it is quite unlikely that such changes would give rise to intact spiritual beings.

In that a society of spirits are seen in NDE's and in that reincarnation requires insertion of an outside entity, random evolutionary changes cannot be considered as possibly giving rise to a human souls. There is also a mechanism that transports souls to a spiritual realm. Random occurrences do not account for this.

A Holographic Universe

This topic is explored in depth in the following chapter. In summary it is concluded that our world is not a hologram.

A Computer Generated Universe

This concept utilizes an intellectual construct to make assumptions that have no evidential base. It is discussed later in the text. In summary it is concluded that we are not part of a computer program.

A Technologically Advanced Species

This hypothesis assumes that a technologically advanced species will have the ability to create and dwell in a spiritual realm. Also, not only will they be able to transfer individual identities into and out of this realm, but they will also be capable of re-experiencing human life: at will! This hypothesis not only accounts for what we know, it also suggests how future technology available to an advanced species could have been used to accomplish this and why it was applied.

Although challenging to adopt, this concept offers a stepwise approach toward determining the nature of spiritual existence. In summary it is concluded that this explanation is logical, is supported by what is known and represents a reasoned approach to this issue. Is it correct? The future will have to determine that.

Chapter Fifteen

The Holographic Universe

Do we live in a hologram? Michael Coleman Talbot's book *(A Holographic Universe (first published in April, 1991 by Harper Collins ISBN 0-06-092258-3)*, considered and supported the possibility that the universe is holographic in nature. That is what we perceive as reality, such as a tree or a rose, is a holographic image interpreted by our brain as being a tree, or a rose.

His opinion is based on the work of many scientists, but most notably on the evidence presented by Dr. David Bohm (1917 to 1992), Dr. Karl H. Pibram (1919- to date) and Dr. Karl S. Lashley (1890-1958).

David Bohm was an outstanding physicist very involved in the evolution of quantum mechanisms and the wave nature of electrons. He was fascinated with the double slit experiment that demonstrated that electrons had a dual nature; they could act as waves forming interference patterns when passing through the slits in a device or as particles when the beams passing through any given slit is observed and measured. He also was fascinated with the Heisenberg uncertainty principle where it was found that the exact location and angular motion of electrons could not be completely calculated simultaneously. If one was pinpointed, the other was impossible to measure accurately. These phenomena led him to develop his concept of "Implicit Order" in physics that was based on his belief that all energy and matter were somehow an intimately connected whole. Concisely, David Bohm's implicit order theory proposed that reality as a construct of separate particles is incorrect and that basically reality consists of a single series of connected electromagnetic waves.

Enter Dr. Pribram who early in his career was a student of Dr. Wilder G.Penfield, a highly regarded neurologist. Dr. Penfield was particularly noted for his experiments to determine the location of memories in the brain. He discovered that by stimulating the temporal lobes of epileptic patients undergoing surgery, he could elicit in them, memories of past events in great detail. He therefore concluded that specific memories were stored in specific brain loci; that he named engrams.

(However, Dr. Penfields' findings were not correct; it was later demonstrated that they resulted from faults in the methodology he used. Additional studies demonstrated that reactions similar to those he observed only occurred in epileptic patients).

Key evidence pointing to the fact that Dr. Penfields' conclusions were incorrect was established by Dr. Karl Lashley who initially performed his experiments on rats in an attempt to confirm and expand on Dr. Penfields' findings. His results however proved to be very different.

The method used by Dr. Lashly (1929-1950) in his experiments was to train rats to run a maze, surgically remove part of their frontal lobes and then check on their post-surgical ability to run a maze. What he found was that his trained rats could still run the maze regardless of which section, or how much of their frontal lobe he had removed. They were a bit slower and slightly more uncertain as a function of how much of their frontal lobes had been removed, but they still ran the maze faster and more accurately than untrained rats. Their memory had been maintained.

In 1969, when Dr. Pribram learned about Dr.

Lashley's studies he was amazed. He then noticed that brain-injured patients who had large sections of their brain removed did not lose their memories but they did however become less and less clear if more and more brain tissue was removed. As a result of these findings Dr. Pribram concluded that memories were distributed throughout the brain and were not localized in particular brain cells. At the time, however, there was no concept to define how this could occur. But in the mid 1960's, he learned about laser holography after reading an article in the Scientific American.

He learned that holographs could be created either by using one or two lasers. By way of example, one can use one laser to illuminate an item, such as a rose, and have reflected light from it impinge upon a photographic plate (a negative), while the other laser is aimed directly at the photographic plate. The two beams of light striking the plate cause interference patterns to appear on the plate (negative). If you look at the photographic negative following this exposure all you would see is a blurry, seemingly meaningless image. However if you pass a laser beam through the negative you will see a 3 dimensional image of the rose.

Interesting, but a further step is more than interesting, it is fascinating. That is if you cut the holographic negative in two you would see 2 complete images of the rose. And if you cut it any number of times you would see an equivalent number of smaller, but complete images of the rose. Additionally, the cut upper right edge of the hologram negative would present the image as if it were viewed from the upper right and if cut on the lower left the image would appear as if viewed from that perspective.

This phenomenon induced Pribram to firmly believe that this provided him with the mechanism responsible for the diffuse nature of memory that had previously eluded him. He therefore postulated that memory developed in a similar holographic manner, with interference wave patterns generally impressed and stored throughout the frontal lobe. Pribram also noted the similarity of his concept of holographic memory to Bohms' idea of Implicit Order and suggested that they cooperate in promoting his hypothesis that memories are distributed throughout the brain as fixed interference patterns similar to those that occur in holograms. He named this his *holonomic* concept of memory accretion and storage.

Essentially this model of cognition centered on the fixation of wave interference patterns resulting from the firing of a number of dendrites, thus establishing fixed, but retrievable, specific memories. This concept has thus far received considerable support and confirmation from additional studies and the scientific community.

Another theory proposed for visualizing reality and the existence of consciousness is that proposed by Hameroff and Penrose *(Hameroff, S.(1987) Ultimate Computing, Elsevier) and (Penrose, R.(1989) The Emperor's New Mind Oxford University Press)*. They concluded that the numberless microtubules in each of a series of neurons can open and close with extreme rapidity and give rise to a continuous flow of information (much like a computer does with zeros and ones) that we perceive as consciousness. Referring to paramecia again, Dr. Hameroff noted that they do not have neurons but do have microtubules and do demonstrate cognitive capabilities (see page 22). This observation was key to his concept that microtubules are the basis for cognition.

One can easily visualize the "firing" of microtubules giving rise to consciousness while undoubtedly also forming quantum waves in the process. These quantum waves then in turn give rise to interference patterns that are fixed in the brain thereby establishing memory. Clearly, I believe that their theory does not refute the quantum wave/particle theory presented here for establishing memory and in fact supplements it. An additional point of interest is that Dr.

Hameroff believes that information defining Self survives mortal death as entangled quantum packets. *This is what I believe as well!*

As I am certain you realize, this is a complex issue far from being completely presented here. For example, which are the areas of the brain that are involved in acquiring and storing information and what about the difference between storing short term and long term memory? Complex indeed, much more so then we will or can factually explore here.

Moving on, another theory proposes that interference properties also relate to auditory functions. This concept has been proffered by an Argentinean-Italian researcher, Hugo Zucarelli, but although he has produced recordings based on this concept that reproduce sounds with extreme accuracy, his 'holophonic' concept remains controversial.

Now let us go back to the proposition proffered in Michael Coleman Talbots' book "*A Holographic Universe.*" Mr. Talbot extrapolated from the information he provided that everything in our universe is holographic in nature. He further postulated that what our minds interpret as material objects are in actuality holographic interference patterns emitted by waves transmitted from specific quantum stimuli existent in a truer, deeper, reality.

On the Dual Reality of Existence

In essence he concluded that since the brain uses holographic information that in turn is visualized and interpreted by us as an image, that the brain itself is a hologram. So then if our car uses gasoline to run and we all know this, is our car really gasoline!! At a sub-microscopic level all our tissues can be visualized as strings or quanta but this in no way indicates that our brains are visualizing holograms. What is missing in the theory of a holographic universe is one simple concept; this concept is known to us as common sense. I am certain that all of us who fly in planes just know that we are not flying holographic waves. If we hunt and kill a deer, it is dead. If we dine on venison and find it delicious that is not a hologram-matic experience. Let's face the obvious; we do not eat holograms, we eat food. I eat, therefore I am. The reality we perceive at the level we perceive it is true!

So much for that concept and the further notion that our existence is a holographic illusion. Reality is, as we see it, and as we sense it. That we visualize it by a system (the brain) that uses a method similar to holography to experience consciousness and retain memories in no way alters reality. Holographs of objects projected in 3 dimensions are exact replicas of the real objects they were created from, not the other way around. The interference wave patterns are, and represent, intermediate templates useful for the reconstruction of reality, much as a mold is used to create a final form.

If we wish to think that we are a conglomerate of cells and fluids consisting of molecules, atoms, subatomic particles, waves or intertwined strings, all good and well. This, however does not preclude our existence as living beings nor does it preclude the existence of reality at the level we perceive it.

Conclusions

I believe that consciousness is produced by the firing of microtubules in neurons as proposed by Dr. Hameroff and that memories (as well as other phenomena to be addressed later) are kept in the brain as captured, modified and fixed interference waves. Living humans are not nor will they ever be holograms and do not dwell in a holographic universe as proposed by Coleman Talbot.

Chapter Sixteen
_____

The Simulation Hypothesis

Nick Bostrom, an eminent philosopher, has put forth a number of concepts of interest and currently under discussion. But in my opinion, his most interesting and controversial concept proposed in 2003 is his simulation hypothesis **("Are you Living in a Computer Simulation?", Philosophical Quarterly vol. 53 (2003), No 211, p. 243-255.)**

In it he proposes that there is a significant possibility that reality (1 chance of 3), as we experience it, is the construct of an incredibly powerful computer generated by a post human intelligence/society. The possibility that more than one computer is generating more than one simulation is not ruled out. In essence his philosophical concept is based upon 3 potential premises of which only one is likely to be true.

These are first, that very few, if any, societies will achieve a post human level (we would probably become extinct before this occurred). Second, if there were any survivors, few would be interested or wealthy enough to study ancestor simulations and finally, if the proposed third case were true, most of us are living in a simulation and the probability of this being so is close to being absolute.

This is a fascinating concept that denies our logic and sense of actually existing in complete reality. The theme can be visualized much as in the film "The Matrix" except that an evil controlling entity is not part of this thesis.

There are several difficulties with his concept (and undoubtedly more that I cannot think of at this moment) that I must take issue with. His proposition is based on 3 premises. First, the assumption is made that there are no other options; for example, however, what if many human civilizations attain posthuman capabilities? What if such civilizations have options other than computers or would prefer to experience ACTUAL existence (more on this later)? What if superior post-humans have better things to do than run what we would now consider extremely sophisticated simulations? What if computers, no matter how powerful, cannot construct a universe exactly like ours? What if the true nature of quantum uncertainty cannot be computed with certainty? Why preclude the possibility that the existence we experience now is real; isn't that most likely? Does this mean that a superhuman society cannot evolve? If you are walking in a field and suddenly come upon a snake do you think it likely that this experience had been preconceived in a computer program? Why confine proposed future post-humans to only three options? And finally, where is hard evidence confirming this concept?

No, I enjoy the concept and see seeds of my own ideas in it, but I do not accept it anymore than I believe that we are living in a holograph.

Conclusions

We are not computer manifestations. Although we do not understand existence completely, we know that we are born, that we live and that we die in true reality. We know our history, our biology, our chemistry, our environment, and much about our universe. The essential questions are 1) what the complete nature of true reality is and 2) what happens when our physical bodies die?

Chapter Seventeen

Dual Reality

We know so little and we have so much to learn, nonetheless it is surprising how much progress we have made and are continuing to make and at an accelerating pace. Our advances in medicine, conveniences, weapons, life extension, understanding our universe, entertainment, and communication would seem miraculous a hundred years ago. We are no longer old at age 65; in fact, the majority of us may live to be centenarians.

We *(Dr.Philipp Holliger et al of the UK Medical Research Council's Laboratory of Molecular Biology)* have already made synthetic versions of DNA and RNA that mimic the chemistry of life and are exploring the possibility of a future synthetic biology. Modifications have even been made that can pair with natural DNA. Imagine the possibilities that could develop from this.

Additional studies also consider the possibility of other chemical structures with similar capabilities. Such potential variability greatly supports the possibility that life exists elsewhere in the universe. Clearly, our intimate knowledge of the biology of life is rapidly growing and approaching the extraordinary.

It is very likely that in the next hundred years or so, it will be possible to re-grow amputated body parts (we replace hearts, lungs, kidneys and livers now), live for an indefinitely long number of years and perfect each and every parameter of our biological existence to as close to utopian perfection as possible – civilizations' idiosyncrasies permitting.

But what if this has already happened – somewhere else in another galaxy, somewhere in our galaxy, somewhere in our solar system or perhaps even here? The universe is vast and the criteria required for life have, (much more likely than not), occurred in a fashion similar to ours on other planets. So then, what if it has indeed already happened but we are just not aware of it?

Conceive of this: an incredible absolutely perfect society. Everything that could be accomplished has been accomplished. Living there is just short of being completely utopian. There are few, but surmountable, negatives, but immortality and a perfect society have been achieved. Our future superior human (or future alien) descendents have created perfection. But we are getting ahead of ourselves. Let's review the evidence we have accrued thus far:

1- Out of Body Experiences

The OBE of Pamela Reynolds was confirmed in that her surgeon verified that she had correctly identified the Midas Rex bone saw he used to open her skull. In the other case cited, Ms. Clark confirmed that Maria, in cardiac arrest, envisioned and identified in detail a shoe on a distant window ledge. Although not a science that can be repeated at will, these confirmed reports go a long way toward supporting the existence of an internal spiritual essence that is separate and distinct from our conscious carnate mind. Succinctly then, we seem to encompass two distinct realms, material and spiritual.

2- Reincarnation

The existence of reincarnation has also been confirmed. This provides evidence that the essence (spiritual manifestation) and memories of a deceased individual can survive physical death. Again this confirms the coexistence of a spiritual essence that separates from the body on mortal death.

3- Near Death Experiences

Also confirmed, but additionally pointing to the existence of spiritual entities in a spiritual society. In NDEs this existent spiritual world is visualized such that in all ways it is sensed as being both real and ideal. Remember, near death experiences have been confirmed; as such, the visualization of a spiritual realm is not to be dismissed.

We propose, much as Bostrom did, that in the future superior humans will exist possessing currently unimaginable technology. Consider this: quantum teleportation, using quantum entanglement, already exists. Anton Zeilinger et al transmitted the exact information of an entangled quantum particle (limited by the speed of light) for a distance of 89 miles without the actual particle having traveled. (*"Qeuantum teleportation using active feed-forward between two Canary Islands." Anton Zeilinger ArXiv.org 5/17/2012."*) Startling advances such as this demonstrate the accelerating pace of technological development (and could possibly be developed into the means for gestalt transfer). As a result of this, humans will eventually realize the advantage of merging with machines. The marriage of humans with machines is not only conceivable and probable, it is inevitable. Ray Kurzweil's predicted singularity is near!!!

Clearly, bodies that eventually do malfunction are a liability. They will not allow immortality. Thus, existence will inevitably first be transferred to machines and then to the equivalent of living interference patterns, or specific quantum packets encompassing each entities individuality. This sequence of events would account for the spiritual world that evidence indicates exists. It would seem that these entities can shift their appearance and appear to be angels, deceased relatives, or any other ethereal form they wish.

The first question to be addressed in regard to these experiences is HOW they could come to be. Clearly, we do not know exactly how this happened but we can hypothesize. If we base our predictions on a future where superior humans can accomplish what would seem miraculous or impossible now (much as our ancestors would have thought about our current capabilities), perhaps we can determine HOW all this has occurred.

So, from this point forward we must stretch our imaginations to account for a seemingly incredible future. As you know we have already developed electronic games that we can control. Soon they will be in 3D. At the same time science is considering the possibility of transferring the persona of one individual to another. This is just in its infancy, but in all probability it will eventually be a reality much as depicted in the movie Avatar, but scientifically based and accomplished.

Right now, for instance, individuals can currently "be" in 2 places at the same time. An example here would be the work of **Mel Slater, Professor of Virtual Environments at University College London (UCL).** He and his team have "projected" people from Barcelona to London as robots; providing each (the robot and the human) with mechanical means for joint motion, hearing devices for sound, T.V. receivers/cameras for eyes, and vibratory sensors for the sensation of touch. The human in Barcelona then experienced being in London by being "wired" to the robot there.

In such a manner you could be in New York and experience that you are in Paris. You would see Paris, you could see and talk with people in Paris, and you could walk there, and so on. This is not the future. This has been done (Barcelona to London). Primitive? Yes, but so was the Wright Brothers first aircraft.

But how could we conceive of having the spiritual essence of an individual transfer to another person? Well, with a little imagination we may be able to construct a device that could do just that.

Let's go back to Dr. Pribram and his widely accepted assertion that memory is dispersed throughout the brain as interference patterns in a holographic manner and look at where we are now. For starters, let's examine the workings of electroencephalograms, pet scans and MRI's (magnetic resonance imaging). EEGs record brain waves as electronic impulses that can be seen on paper or on a computer screen as wavy lines. It detects alpha waves, beta waves, delta waves and theta waves providing a rather complete picture of the electrical activity of the brain.

MRI's create images using magnetic and radio waves. These powerful impulses affect the body's atoms forcing their nuclei into a different position. As their nuclei move back in place they emit radio waves of their own. These signals are picked up on a scanner and are turned into a picture on a computer.

Pet scans (positron-emitting radionuclide scans) are based on tagging specific compounds with radioactive materials that will act as tracers, localizing in specific organs of interest to a physician. The radioactive emissions are captured to provide an image of the particular organ of interest. At any rate, I do not believe that this particular technology will be of use for spiritual transfers. CAT scans make use of x-rays (radiation) and are therefore also unlikely to be used for this purpose.

On the Dual Reality of Existence

Future devices that will essentially be variants of EEGs and MRIs in combination or with lasers or other advanced devices may provide the mechanism required to produce interference patterns in a recipient's brain. I can visualize initial tests involving attempts to transfer memory in rats. In this case, rats would be wired with electronic sensors and a transmitter to collect their brain wave patterns as they are taught to run a maze for the reward of a morsel of food. Attempts would then be made to transfer this information to the brains of untrained rats. Successful transfer, without any indication of damage to the recipient, should be easily recognized. If it succeeds it will undoubtedly be recognized as clearly being a Nobel Prize level achievement!

(Farfetched? Not at all. At the time of this writing scientists at Stanford University are working on a device that receives brain waves, transfers them to a computer, interprets them and converts them to words. Stephen Hawking is currently cooperating in this effort to allow him to communicate clearly.)

The next step in memory transmission would involve the transfer of a memory (for example a single image) in humans; such as implanting the memory of seeing a particular vase from one individual, say in Minnesota, to one or several individuals in Madrid and or Thailand. If this experiment were successful it would totally revolutionize learning and would undoubtedly lead to the downloading of huge amounts of information from computers to the brains of humans. In a natural follow up to this, the patterns of humans will be downloaded to quantum computers so as to take advantage (meld with) of their incredible knowledge base and fantastic processing speeds). Clearly these sentient computer/humans would incredibly speed up our development and understanding of the universe and reality. Also, machines can be repaired so at this point immortality, for the most part, will have been achieved. True immortality, however, will only be arrived at when transfer of self to a nonmaterial, spiritual existence has been realized. This final step should be greatly accelerated as a result of the merging of man with quantum computers.

Therefore, melding of future superior humans to super computers would just represent the first step to be taken. It could occur at a point near human death or at a selected point in human life. "Donor" individuals submitting to gestalt transfer to a supercomputer would not be altered, but would remain intact as functional mortal beings. All aspects of their melded counterparts, however, would essentially exist forever with transcendent cognitive capabilities.

Following this, will be the ability of individuals to transfer their entire existence to a quantum (spiritual) reality. This potential method is not as unreal as it may sound. We will shortly have quantum computers that will store and very rapidly retrieve huge amounts of information as quantum bits, and quantum bits are incredibly small. Eventually all the data required to define being alive will be stored as entangled quantum bits.

The next step for achieving complete immortality would be to transfer "self" from quantum computers to freely existing and specifically configured entangled quantum bits. This incredibly small packet of information defining self will be immortal, considered to be spiritual and a true avatar. (As noted, this is how Dr. Hameroff believes that information defining self survives mortal death (see pg. 95)). There is no reason why these procedures in reverse could not allow for the transfer of ones gestalt (being) from the spiritual state to an existing mortal state; a further avatar-type transformation.

WHY such a transfer would be desirable will be discussed shortly, but in this case it would be essential to separate and mask the spirit of the donors (gestalt pre-configured quanta) from the conciousness of the recipients. This to allow the transferred "spirits" to

On the Dual Reality of Existence

completely live the experiences of individuals they have been transferred to without being aware of their own spiritual existence. (In this sense those experiencing reincarnation, by recalling their past identities, would have had a somewhat failed spiritual transfer; the main point being however is that reincarnation proves that a deceased persons entire being can be transferred to a living being. It has been done – many times –and is still happening!!).

The final step to be mastered would be retrieval of the inserted spiritual entity on the death of its mortal host. This could be a simple function of monitoring the available oxygen level in a host's brain or by any number of other straightforward methods. This also has been done – has been proven to have been done – and is still being done!

I admit that this concept requires a complete reconsideration of the nature of reality, but as far as I am aware, this transfer of souls is actually happening now. If it is not too difficult to conceive of us as consisting of strings or believing that every time we make a decision we split our existence between our existing universe and a newly created universe, then conceiving of quantum transfers is not that exceptional and we do have the confirmed evidence of OBE's, NDEs and reincarnation.

There are many incredible beliefs that are now accepted or are being considered. Clearly, however, the dual reality thesis presented here is more probable than a proposed reality of living in a hologram, or being computer constructs in a super human created computer game, or any other proposed but unsupported cosmological surmises.

So then, based on the projection of a superior civilizations' progress and with a fact-based extension of what currently has been proven to occur (OBE's, NDE's, reincarnation) we have accounted for spiritual existence and the transfer of individual spirits into individual living (as we know living) humans. We have proposed HOW all this can be accomplished, but not WHY it would be done. And WHY would such transfer be masked from both the gestalt donor and the gestalt recipient; WHY would this be desirable for immortal spirit entities?

Consider this, what if it is important that we do not know what occurs after physical death; this, not to prevent fear, but rather to prevent knowledge. What if our bodies are receivers and our spiritual selves our immortal essence? What if knowledge of the exact nature of our existence after death would lessen our carnate life experience? What if our incarnated life experience is in a sense the salvation of our

On the Dual Reality of Existence

existence after death? I will tell you then what we escape by temporarily re-existing as mortals!

Let us use Occam's razor, which as you know in essence states that when you have competing theories the simplest one is the best. So what then could be the simplest possible reason for immortal beings wanting to re-experience life?

The simplest answer is because they escape from what is the worst fate of all: eternal boredom. How can this be? Eternal boredom, why is that? What I am stating is that immortality, unmodified, gives rise to boredom, the latter being a significant drawback to an ultimately advanced immortal race and society. Again, living forever unavoidably induces boredom and boredom leads to depression. Consider what you would do with yourself if you have forever to keep on doing it? I cannot conceive of existing in a society where everything that could be done has been done and where everything you would want to do you have already done numberless times. This is not an eternity anyone would particularly desire; but fortunately, however, this problem was solved.

It was determined that the only answer to eternal boredom is mortal life, not the creation of an advanced computer program creating a matrix mimicking life on a quantum computer, but sensate, unpredictable, twisting, forming, dissolving, struggling, exciting, dangerous, rewarding, frightening, joyous and ending mortal life.

Again, spirit memories of past mortal lives, such as eating a delicious steak, or running on a beach, breathing, feeling the weight of gravity while relaxing in bed, enjoying sex, experiencing physical strife, and challenging the unknown, the remembrance of temporal existence, is not the equivalent of literally being a living human. The unpredictability of mortal physical existence can in no way be substituted for by memories.

Perfection and immortality have their costs. Spiritual beings may obtain immortality, but nonetheless at the cost of losing most of their past animal sensate pleasures. But with advanced knowledge and technology the essence of spiritual beings can coexist with humans. Why humans? Because humans have fascinating capabilities and limitations. They are relatively intelligent, create coherent societies and are so configured as to accomplish many things. Most significantly their life experiences are fascinating, they feel pleasure, enjoy physical reality, appreciate nature, are mortal, and they die. *Thus, in the end, by being mortal they free the transferred spirit to return to eternity.*

Transferring a concealed identity into a human

On the Dual Reality of Existence

should not be difficult for an advanced race where science and knowledge is in essence complete. Even now we have genetic engineering where we can insert genes from one living being to another and from one living species to another. Right now we have almost instant communications, GPS devices, medical miracles, and other engineering marvels such as heating and cooling devices, aircraft, radio, television and so on. We have conceived of nuclear energy, quantum mechanics, relativity, dark matter, and multiverses. We understand evolution, black holes, and the very fabric of space-time.

Think then of an evolved species existing eons before intelligent live appeared on earth. Think what they could have accomplished. From supersonic travel to instantaneous quantum teleportation travel, from mortality to immortality, from transferring genes to transferring the essence of individuality. From immortality to temporary mortality? How difficult would it be to direct the essence of a persona to a newborn creature, a newborn human? Would humanity have had to be modified to receive an external essence? If this were so would this be difficult for a race with ultimate knowledge? I do not believe so. A "simple" self-replicating receptor modification could have been inserted early in the evolution of man. And as the inserted identity is of a spirit nature, there is no problem of a residual body to store when it is cohabitating with a human. Insertion of the individual spirit could occur in utero or at birth. Advanced location devices would direct specific spirit identities to the newly conceived. The inserted being would evolve with a sense of self that is identical to that being experienced by the newly born infant. The adventure would thus begin.

Now you have learned WHY an immortal being would want to re-experience mortal life. You have learned why they would wish to escape perfection to enter a known and dangerous reality. The answer again is simple and direct, to escape boredom. In no way can a utopian society duplicate the complex and unpredictable nature of life, its experiences, its exhilarating successes, its dark failures, or its everyday existence of uncertainty. *Life offers the unknown, the stimulation of an uncertain outcome.*

And uncertainty does not exist in a utopian society; everything is assured. Immortality without negativity, literal perfection forever initially sounds wonderful, but on examination it's disquieting. Within itself immortality can never present the stimulation of uncertainty. Thus mortal actuality, harsh reality, is the answer to certain, uninterrupted and totally predictable and unavoidably boring eternal bliss. It

On the Dual Reality of Existence

would also not be surprising to learn that such advanced beings are living among us now, maintaining their original functional memories, so as to actively study our civilization.

In this temporary transfer to human existence, no matter how awful such a life experience turns out to be, at its end a return to a utopian existence is assured. Thus the unending bliss of utopian immortality can be periodically spiced with unpredictable mortal life. Has this been done, is it being done? No one knows for certain, but if this is indeed what is happening, it would go a long way toward explaining currently unexplainable "paranormal" experiences.

So if you were living in a utopia forever, you could escape crippling boredom at any time by experiencing corporal transfer. This as a form of entertainment and a practical way to escape boredom, much as we now go to the movies or watch television to escape being bored. In this proposed utopian spiritual existence, the temporary transfer of self to humans has now been accomplished. In this realm you simply transfer your persona to a living existing species in this or in another world where you deliberately do not know, or want to know, of your immortality or your fate - be it early death, fame, misfortune, hunger, power or any other possibility.

It also may also be possible for you to select the

nature of your new mortal experience. You may choose to be white or black, rich or poor, Muslim or Jewish but this would lessen the degree of uncertainty basically desired. For example, if it is possible (and I am by no means certain that it will be), you may choose to be an eagle or some other living creature; but in this case as with any other selection, to maintain unpredictability you would, here again, not know its fate. Directly put, to truly defeat boredom unpredictability is absolutely essential.

How long would an experience be? That will depend upon its outcome, anywhere from minutes to years. Assuming however an experience interval equivalent to what we have now, a most probable maximum time allowed per life experience would be about 100 years: the approximate lifetime of the selected animal, man. By selecting an undetermined life outcome, what occurs during that period would depend upon your random birth circumstances and would vary from there as a function of an unpredictable future. Clearly one could not be bored under these circumstances. Depressed, yes, occasionally, but truly bored, very unlikely.

Thus by transferring your gestalt being to a mortal the final negative aspects of immortality are defeated. Coexist with your human mortal host for at most about 100 years, leave and return to your immortal utopian existence, be greeted by your loved ones and review what you have experienced. Re-experience utopia for several days or several eons and when desired, return to another human host for a new, unpredictable, exciting and never ever truly boring interlude. With immortality assured and boredom conquered, true utopia will have been achieved.

And think of this, in infant James Leininger's nightmares he would exclaim *"Little man can't get out,"* not "I CAN'T GET OUT"! A superior transplanted soul would say exactly that! The expression *"little man"* used referring to the relatively simple human body the superior soul was experiencing. A coincidental turn of phrase; possibly, or perhaps startling evidence pointing to the dual reality of existence.

Conclusions

For an immortal society possessing ultimate technological capabilities, gestalt (quantum spirit) transfer to other beings will be (and probably has been and is) both possible and desirable. If this is the case, true utopia will have been achieved.

Chapter Eighteen

Other Possibilities

Although the following premises give rise to disturbing and somewhat confusing possibilities, they must be noted.

There are other possibilities. Whereas the first possibility, already described, is one where the spiritual soul maintains a passive role much as in watching television or a movie; the second possibility is that of assuming an active role. In this latter case the spirit itself is in control of its adopted mortal body but not retaining memory of its spiritual essence.

There would be no particular technical difficulty in establishing this type of transfer using methods similar, but slightly modified from that used for passive transfer. What is more significant here is that the activities of the mortal host are totally directed by its spiritual guest. This raises a question of accountability for which I have no answer.

I assume, however, that if this controlled mortal proved to be evil, its spiritual controller would necessarily be accountable. Any judgment passed could range from minor to severe dependent upon the nature and duration of the transgressions. Therefore, such a transfer would be risky and less desirable than a passive transfer. (It is interesting to note that in all reported cases of reincarnation a passive role for the transferred soul is noted).

A third possibility is transfer while maintaining complete memory of the true nature of their spiritual existence and would more than likely be used to study current human culture. Any transfers of this type, however, would require strict screening to eliminate any possibility of influencing the ongoing development of human society. This restriction would hold true unless there was a compelling agreed upon reason to initiate a societal change.

The final possibility, already touched upon, is a transfer through time; this for adventure or for study.

Conclusions

On the matter of time manipulation, if it should become available, there is no reason why it would not be used. Full memory-intact transfer would only be used for scholarly reasons and would require candidates to undergo and pass a complete personal attribute evaluation. This method of transfer would be strictly limited and carefully used.

Chapter Nineteen

Remaining Questions

Are there other cultures in space? In this text we have described a superior culture utilizing humans essentially for their amusement. It is unlikely; however, that it is the only advanced culture in a universe with 100 billion galaxies and with about 100 billion stars on average in each galaxy.

Here, UFO's provide proof of the existence of one or more advanced cultures. One question now is how many of these cultures are far from being benign? Surmising that there are hundreds if not thousands of "alien" cultures in the universe it is very likely that some are aggressive and could be extremely dangerous. In fact claims have been made that spacecraft have been visualized firing at each other in our upper atmosphere. On the other hand some cultures could not only be benign but also be protective. Clearly, we are babes in the woods and have far more to learn about the nature of our universe than we know.

Another critical question that has not been addressed to this point is: who or what has died and who or what has entered the spirit realm. The basic question here is whether the spirit leaving the body is that of the deceased individual or if the departing spirit, (now containing the essence of the deceased individual) is a returning transplant from an advanced race.

Evidence from reincarnation supports the fact that separate and distinct spiritual entities can coexist with living beings. If this is indeed the case it is likely that humans were selected as carriers by a superior race. Again then the question becomes who has died?

Assuming the existence of a transmitted soul, what, (not who) has died is the vehicle supporting their experience of life. This can be envisioned as being equivalent for them as driving a car or riding a horse is to us. Most important here is that the driver is not the physical vehicle.

On the other hand one may assume that the soul is an innate part of humans. This hypothesis, as any other, must be justified by evidence. One can claim that OBEs and NDEs provide such evidence and possibly that is the case. However WHY humans should possess souls is not answered. There are religious reasons but not evidential reasons. Humans are mammals; do all mammals have souls, do rats have souls? Again, recall that reincarnated spirits exist and that they are separate and distinct from their hosts.

Also, we know that souls exist and we also know that an ultimately advanced society would have a reason WHY to project souls into humans.

So I believe that the human, who has died, was the vehicle and what is actually the transcendent 'you' survives, all experiences from every visitation made, maintained.

In NDEs 'you' transfer to a spiritual realm and see deceased loved ones, but as has been pointed out this in all probability, is part of a process of reorientation. The loved ones seen in an NDE may have actually existed as family and friends in their last incarnation, but in actuality they are fellow spiritual travelers. In no way can I pretend to know the relationships that exist in the spiritual realm, but I assume that they are, in essence, perfect.

Another intriguing question is whether all human potential hosts are used by utopian gestalt souls. Interesting indeed! This question remains unanswered but I believe that it is probably unlikely that every human has a gestalt soul. These unfortunate humans, without gestalt souls, will in every sense be mortal and will not survive death. Do they in any other sense differ? Do they, being "unmonitored", account for the more brutal aspects of life? Possibly, but this is speculation, not deduction.

On still another point, is it possible that utopian beings could literally die due to a botched transfer or some other unfortunate occurrence? I believe this can occur since transfers are based on a specific configuration manifested in the recipients' brain. For this reason a gunshot wound to a critical region of the brain could quite possibly destroy the transfer site before transfer could be effectuated, thus destroying the transferred entity instantaneously. It may also be possible that a poison causing instant toxicity could alter such site preventing spiritual transfer.

Finally, again, what about God? How does a supreme being enter into this equation? Nothing written here has supported or was meant to support or negate the existence of a supreme being. What has been questioned is the nature of the soul, far removed from the proposed existence of an initial creator of all matter and energy. But then, what about God?

On the Dual Reality of Existence

As noted before, nfortunately, there are unquestioned difficulties relating to the existence of a supreme being. Why are their different and conflicting religious beliefs? Again, why have the last significant proponents believed to have presented evidence of such a being lived 1500 to 3000 years ago? Why would such a being permit genocide, pain, pestilence, war, crime and hunger? A supreme being would not permit this. Religious claims that we are being tested have a shallow, hollow, ring.

It is simply impossible for us to pursue this question. The Kaballa states (quite accurately) that only a fool would waste their time trying to understand God. As previously stated, here we must live without evidence except for flowers, air and water.

Conclusions

Although there are myriad possibilities that could define reality, verifiable evidence indicates that a dual reality of existence is reality.

Chapter Twenty

Our Fate

Have you noticed? The hypothesis presented for the development and existence of individual gestalt quantum souls in a spiritual realm has been based upon their assumed technological evolution from a physical reality to a spiritual reality. Proposed here was a stepwise manner by which this could have been accomplished, extrapolating by example from the technological progress we down to earth humans are making right now and will continue to make.

It is therefore quite conceivable that the end point of our technological evolution will also be immortality in a spiritual utopian society (or has this already occurred; think about it).

What will happen then? At that point we will not host gestalt guests. Our current visitors will be forced to look elsewhere for subjects to coexist with unless time travel has been mastered. If this were the case, utopian gestalt spirits could transfer to any past period, thus avoiding the loss of human recipients.

However, regardless of any readjustments required by our current gestalt guests, when our civilization becomes immortal and utopian will we then also seek donor recipients? I believe that this is certain.

If time travel had been mastered, we also, might very well transfer our gestalt beings to humans existing in the past. Nonetheless, true utopia will have been achieved again, this time for us.

Conclusions

Many proposed answers have been provided regarding the nature of mortality and immortality thus far, but by all means, not all the questions that could have been asked have been asked and not all questions that have been asked have been unequivocally answered. Only the future will provide total and undeniable absolute truth.

Addendum

The thesis presented in this text is based upon established fact, advances in technology, and assumptive hypotheses relevant to the existence of a dual reality. For this reason many peripheral so called paranormal events were not discussed. Covering these materials would have mislead readers into assuming that the text is concerned with the general topic of paranormal phenomena; this is not its intent and it does not.

However so as to indicate that the remaining paranormal topics were not overlooked, a brief synopsis and my opinion relative to these occurrences will be presented here. Please consider these to be completely separate and distinct from the preceding text as there are no proofs and there is no technology involved. What is presented below is simply my opinion on these reported phenomena.

Crypids

Crypids are animals believed to exist, such as Big Foot or the Loch Ness Monster that have not been confirmed by firm scientific evidence. Many claimed sightings of such creatures have been made but again, none as yet have been confirmed. These claims are not related to the topic covered in this text and I have no conclusive opinion relative to them.

Apparitions, Poltergeists and Angels

If apparitions exist at all they reflect either a failed transport to a spirit utopia or the refusal of a newly released spirit to leave ones current identity and proceed toward utopia. Such apparitions (again, assuming their existence) would have the ability to take on human form to a varying degree and could possess telekinetic capabilities (remember Expedito and the tire rims).

Angels, if they exist at all, can represent failed spiritual transport or spirit entities designed to oversee material existence. If the later instance were indeed the case, their intercedence in specific instances would be conceivable.

Futurists/Prophets

Individuals such as Nostradamus and Edgar Casey are noted for and have been given credit for their capability to predict future events. I cannot decisively determine if their predictions were based upon subconsciously developed and reasonably intuitive expectations or that in some manner they were able to foresee future events. In the case of Nostradamus, however, his quatrains were always cloaked in such a manner that they could be applied to any number of events and Edgar Casey's predictions were fraught with both failed conclusions and failed predictions. I believe that these gentlemen were both quite perceptive and also that both had intuitive insights, however, they did not truly see into the future.

Mediums/Ouija Boards

Frankly, I believe that most mediums are frauds and that most reported Ouija Board phenomena are contrived. But are some real? I do not know, but for the sake of the possibility that some are real, here is my assumed explanation. This is doubly hinged and requires the existence of apparitions as well as the validity of these procedures.

Who do mediums and other Ouija Board users contact? If they contact anything at all it is definitely not the departed spirit of loved ones. What they could possibly contact are earth-bound spirits who claim to be loved ones. Here it would be the luck of the draw; namely which spirit they have made contact with. The motive of contacted spirits could be benign, so as to comfort questioning individuals, or it could be evil, misleading, or even possibly lead to an attempt to occupy and control a participant. Is this all-possible, yes; is it probable, no.

Satan and Hell

Satan and hell are essentially religious concepts and do not find footing in a dual reality. In the case of passive transfer the human responsible for evil acts no longer exists, it is dead. The spirit on the other hand simply coexisted to experience life, could not interfere, and therefore is blameless. I would also assume that the captured memories of evil individuals could be expunged if so desired, clearing the soul of the returning spirit. The sick, evil person will be totally dead forever.

Could a spirit be held accountable and punished for acts done by an individual? This would only be the case if the spirit were indeed literally the spirit of the deceased individual and would of necessity be predicated upon the existence of a different and judgmental reality.

Supernatural Witches

Utter nonsense.

Levitation

Levitation is the apparent mystical raising of a human body into the air with no means of support. Is this possible? Maybe, but I believe that in most instances it is an illusion accomplished by trickery. However, a claimed witnessed account of levitation by Yogi Pullavar had been reported in 1936 where approximately 150 people saw him levitated from the ground for approximately 4 minutes. Christian saints have also been reported to have performed miraculous feats such as levitation or walking on water. The truth? There is no scientific evidence to confirm these phenomena; however, if they exist at all it is by a mechanism I cannot explain.

Psychokinesis

Psychokinesis is defined as essentially being the movement or altering of objects by thought or by poltergeists. Such movement has been seen and witnessed but in such cases trickery or illusion is the most likely cause and has not been ruled out. On the other hand, I personally believe that psychokinesis is possible, does exist and is achieved by quantum mechanisms not yet understood (again I must refer to Expedito and the tire rims).

At this point I have exhausted my list of purported paranormal events. As you can see, in essence, they can be grouped as essentially unsupported "paranormal" occurrences. Nonetheless, although for the most part they are not relevant to the Dual Reality text they do provide for closure.

Summary

Do you find the concepts of reincarnation and immortality difficult to accept? Well, there are many thousands of reported cases of reincarnation and millions of reported near death experiences. Are they all fantasized? It seems unlikely since, as has been pointed out, independently confirmed cases have been reported. Are these concepts more difficult to accept than that all of us are constructed from strings (the currently accepted scientific theory)? How about all of us living in a computer program, or suddenly springing to life all at once due to the desire to do so by a supreme being? Maybe it's easier to believe that we are all holograms? I don't think so. So then, which of these beliefs seem feasible and which are far fetched?

What would appear to be most logical is that advances in technology eventually will lead to immortality and other incredible developments we can only, at best, surmise. And as our universe consists of hundreds of billions of galaxies, this may well have already occurred. Also, of all the other hypotheses offered, a dual reality of existence best explains the "paranormal" phenomena that have been observed. And finally, proof for the existence and transfer of souls does exist.

Final Conclusion

**Is this the truth? The future will decide.
Have a good day; see you later**

FIN

Check out our website
www.dual-reality.com

www.ingramcontent.com/pod-product-compliance
Lightning Source LLC
Chambersburg PA
CBHW051529170526
45165CB00002B/667